Roger Lee (Ed.)

Software Engineering Research, Management and Applications 2010

Studies in Computational Intelligence, Volume 296

Editor-in-Chief

Prof. Janusz Kacprzyk
Systems Research Institute
Polish Academy of Sciences
ul. Newelska 6
01-447 Warsaw
Poland
E-mail: kacprzyk@ibspan.waw.pl

Further volumes of this series can be found on our
homepage: springer.com

Vol. 275. Dilip Kumar Pratihar and Lakhmi C. Jain (Eds.)
Intelligent Autonomous Systems, 2010
ISBN 978-3-642-11675-9

Vol. 276. Jacek Mańdziuk
*Knowledge-Free and Learning-Based Methods in Intelligent
Game Playing,* 2010
ISBN 978-3-642-11677-3

Vol. 277. Filippo Spagnolo and Benedetto Di Paola (Eds.)
*European and Chinese Cognitive Styles and their Impact on
Teaching Mathematics,* 2010
ISBN 978-3-642-11679-7

Vol. 278. Radomir S. Stankovic and Jaakko Astola
From Boolean Logic to Switching Circuits and Automata, 2010
ISBN 978-3-642-11681-0

Vol. 279. Manolis Wallace, Ioannis E. Anagnostopoulos,
Phivos Mylonas, and Maria Bielikova (Eds.)
Semantics in Adaptive and Personalized Services, 2010
ISBN 978-3-642-11683-4

Vol. 280. Chang Wen Chen, Zhu Li, and Shiguo Lian (Eds.)
*Intelligent Multimedia Communication: Techniques and
Applications,* 2010
ISBN 978-3-642-11685-8

Vol. 281. Robert Babuska and Frans C.A. Groen (Eds.)
Interactive Collaborative Information Systems, 2010
ISBN 978-3-642-11687-2

Vol. 282. Husrev Taha Sencar, Sergio Velastin,
Nikolaos Nikolaidis, and Shiguo Lian (Eds.)
Intelligent Multimedia Analysis for Security
Applications, 2010
ISBN 978-3-642-11754-1

Vol. 283. Ngoc Thanh Nguyen, Radoslaw Katarzyniak, and
Shyi-Ming Chen (Eds.)
Advances in Intelligent Information and Database Systems,
2010
ISBN 978-3-642-12089-3

Vol. 284. Juan R. González, David Alejandro Pelta,
Carlos Cruz, Germán Terrazas, and Natalio Krasnogor (Eds.)
*Nature Inspired Cooperative Strategies for Optimization
(NICSO 2010),* 2010
ISBN 978-3-642-12537-9

Vol. 285. Roberto Cipolla, Sebastiano Battiato, and
Giovanni Maria Farinella (Eds.)
Computer Vision, 2010
ISBN 978-3-642-12847-9

Vol. 286. Zeev Volkovich, Alexander Bolshoy, Valery Kirzhner,
and Zeev Barzily
Genome Clustering, 2010
ISBN 978-3-642-12951-3

Vol. 287. Dan Schonfeld, Caifeng Shan, Dacheng Tao, and
Liang Wang (Eds.)
Video Search and Mining, 2010
ISBN 978-3-642-12899-8

Vol. 288. I-Hsien Ting, Hui-Ju Wu, Tien-Hwa Ho (Eds.)
Mining and Analyzing Social Networks, 2010
ISBN "Pending"

Vol. 289. Anne Håkansson, Ronald Hartung, and
Ngoc Thanh Nguyen (Eds.)
*Agent and Multi-agent Technology for Internet and
Enterprise Systems,* 2010
ISBN "Pending"

Vol. 290. Weiliang Xu and John Bronlund
Mastication Robots, 2010
ISBN "Pending"

Vol. 291. Shimon Whiteson
Adaptive Representations for Reinforcement Learning, 2010
ISBN "Pending"

Vol. 292. Fabrice Guillet, Gilbert Ritschard,
Henri Briand, Djamel A. Zighed (Eds.)
Advances in Knowledge Discovery and Management, 2010
ISBN "Pending"

Vol. 293. Anthony Brabazon, Michael O'Neill, and
Dietmar Maringer (Eds.)
Natural Computing in Computational Finance, 2010
ISBN "Pending"

Vol. 294. Manuel F.M. Barros, Jorge M.C. Guilherme, and
Nuno C.G. Horta
*Analog Circuits and Systems Optimization based on
Evolutionary Computation Techniques,* 2010
ISBN 978-3-642-12345-0

Vol. 295. Roger Lee (Ed.)
*Software Engineering, Artificial Intelligence, Networking and
Parallel/Distributed Computing 2010*
ISBN 978-3-642-13264-3

Vol. 296. Roger Lee (Ed.)
*Software Engineering Research, Management and
Applications 2010*
ISBN 978-3-642-13272-8

Roger Lee (Ed.)

Software Engineering Research, Management and Applications 2010

Guest Editors
Olga Ormandjieva
Alain Abran
Constantinos Constantinides

 Springer

Prof. Roger Lee
Software Engineering &
Information Technology Institute
Computer Science Department
Central Michigan University
Mt. Pleasant, MI 48859, U.S.A.
E-mail: lee1ry@cmich.edu

ISBN 978-3-642-42246-1   ISBN 978-3-642-13273-5 (eBook)

DOI 10.1007/978-3-642-13273-5

Studies in Computational Intelligence   ISSN 1860-949X

Typeset & Cover Design: Scientific Publishing Services Pvt. Ltd., Chennai, India.

Printed on acid-free paper

9 8 7 6 5 4 3 2 1

springer.com

Preface

The purpose of the 8th Conference on Software Engineering Research, Management and Applications (SERA 2010) held on May 24 – 26, 2010 in Montreal, Canada was to bring together researchers and scientists, businessmen and entrepreneurs, teachers and students to discuss the numerous fields of computer science, and to share ideas and information in a meaningful way. Our conference officers selected the best 16 papers from those papers accepted for presentation at the conference in order to publish them in this volume. The papers were chosen based on review scores submitted by members of the program committee, and underwent further rounds of rigorous review.

In Chapter 1, Emil Vassev and Serguei Mokhov discuss their work in creating a Distributed Modular Audio Recognition Framework capable of self-healing using the Autonomic System Specification Language.

In Chapter 2, Yuhong Yan et al. present a new model of the Web Service Composition Problem and propose a reparative method based on planning graphs.

In Chapter 3, Chandan Sarkar et al. explore options for conducting remote usability tests using their newly-developed Total Cost of Administration (TCA) tool to collect and analyze test results.

In Chapter 4, Idir Ait-Sadoune and Yamine Ait-Ameur focus on the formal description, modeling, and validation of web services compositions and suggest a refinement based method that encodes the Business Process Execution Language (BPEL) model's decompositions.

In Chapter 5, Emil Vassev describes his work on code generation of autonomous systems using the Autonomic System Specification Language (ASSL) including an overview of ASSL and features of autonomously generated code.

In Chapter 6, Mohamed Miladi et al. propose a UML extension as a model based solution for the continuous increase in systems complexity and the necessity of cooperation between applications.

In Chapter 7, Haeng-Kon Kim and Roger Lee improve the efficiency of the proxy driving service with the addition of location-based service support.

In Chapter 8, Ahmad Hosseingholizadeh and Abdolreza Abhari propose a new approach to risk analysis by combining various metrics in an attempt to take into account all risky aspects of the software project.

In Chapter 9, Noorulain Khurshid et al. develop a categorical modeling language as the first step toward the creation of a powerful modeling paradigm capable of modeling emerging and evolving behavior of complex software.

In Chapter 10, Vieri Del Bianco et al. investigate the impact of economic factors on the adoption of Open Source Software.

In Chapter 11, Samir Ouchani et al. propose a verification methodology of a composition of UML behavioral diagrams.

In Chapter 12, Haeng-Kon Kim and Sun Myung Hwang address mobile application systems maintenance overhead with a knowledge discovery agent for an effective routing method using simple bit-map topology information.

In Chapter 13, D. Mouheb et al. present an aspect-oriented modeling approach for specifying and integrating security concerns into UML design models.

In Chapter 14, Eric Famutimi et al. present several techniques for using Python as a tool in computational analysis in one dimensional systems.

In Chapter 15, Fracisco Valdés Souto and Alain Abran describe an experiment conducted to compare the performance of the Estimation of Projects in Contexts of Uncertainty (EPCU) model against the Expert Judgement Estimation approach using data from industry projects.

In Chapter 16, Hossein Mehrfard et al. discuss the drawbacks of Extreme Programming (XP) when confronted with the stringent regulations for medical software imposed by the Food and Drug Administration (FDA) and propose an extension to XP to combat its weakness.

It is our sincere hope that this volume provides stimulation and inspiration, and that it will be used as a foundation for works yet to come.

May 2010   Roger Lee
                                                                    Olga Ormandjieva
                                                                         Alain Abran
                                                         Constantinos Constantinides

Contents

List of Contributors

Abdolreza Abhari
Ryerson University, ON, Canada
aabhari@ryerson.ca

Alain Abran
École de Technologie Supérieure
alain.abran@etsmtl.ca

Yamine Ait-Ameur
Laboratory of Applied Computer
Science (LISI-ENSMA), France
yamine@ensma.fr

Othmane Ait Mohamed
Concordia University, QC, Canada
ait@ece.concordia.ca

Idir Ait-Sadoune
Laboratory of Applied Computer
Science (LISI-ENSMA), France
idir.aitsadoune@ensma.fr

Eddy Bell
i-PEI LLC, WA, United States
eddy@i-pei.com

Vieri Del Bianco
University College Dublin, Ireland
viere.delbianco@ucd.ie

Mourad Debbabi
Concoria University, QC, Canada
debbabi@ece.concordia.ca

Khalil Drira
Université de Toulouse, France
khalil@laas.fr

Eric Famutimi
Central Michigan University, MI,
United States
famut1eo@cmich.edu

Abdelwahab Hamou-Lhadj
Concordia University, QC, Canada
abdelw@ece.concordia.ca

Ahmad Hosseingholizadeh
Ryerson University, ON, Canada
ahossein@ryerson.ca

Sun Myung Hwang
Daejeon University, Korea
sunhwang@dju.ac.kr

Mohamed Jmaiel
University of Sfax, Tunisia
Mohamed.jmaiel@
    enis.rnu.tn

Noorulain Khurshid
Concordia University, QC, Canada
N_khursh@
    encs.concordia.ca

Haeng-Kon Kim
Catholic University of Daegu, Korea
hangkon@cu.ac.kr

Stan Klasa
Concordia University, QC,
Canada
Klasa@
    encs.concordia.ca

Dmitri Klementiev
i-PEI LLC, WA, United States
dklem@microsoft.com

Fatma Krichen
University of Sfax, Tunisia
Fatma.krichen@irit.fr

Luigi Lavazza
Università degli Studi dell'
Insubria, Italy
luigi.lavazza@
  uninsubria.it

Roger Lee
Central Michigan University, MI,
United States
lee1ry@cmich.edu

V. Lima
Concordia University, QC, Canada
v_nune@ciise.concordia.ca

Hossein Mehrfard
Concordia University, QC, Canada
H_mehrfa@ece.concordia.ca

Mohamed Nadhmi Miladi
University of Sfax, Tunisia
Mohammednadhmi.miladi@
  isimsf.rnu.tn

Serguei Mokhov
Concordia University, QC, Canada
mokhov@
  cse.concordia.ca

Sandro Morasca
Università degli Studi dell'Insubria,
Italy
sandro.morasca@uninsubria.it

D. Mouheb
Concordia University, QC, Canada
d_mouheb@ciise.concordia.ca

M. Nouh
Concordia University, QC, Canada
m_nouh@ciise.concordia.ca

Olga Ormandjieva
Concordia University, QC, Canada
ormandj@
  encs.concordia.ca

Samir Ouchani
Concordia University, QC, Canada
s_oucha@ece.concordia.ca

Heidar Pirzadeh
Concordia University, QC, Canada
s_pirzad@
  ece.concordia.ca

Pascal Poizat
University of Evry Val d'Essonne,
France
cal.poizat@lri.fr

Makan Pourzandi
Ericsson Canada Inc., QC,
Canada
pourzandi@ericsson.com

Chandan Sarkar
Michigan State University, MI,
United States
sarkarch@msu.edu

Michael Stinson
Central Michigan University, MI,
United States
stins1m@cmich.edu

Candace Soderston
Microsoft Corporation, WA,
United States
csoders@microsoft.com

Francisco Valdés Souto
École de Technologie Supérieure
francisco.valdes@
  spingere.com.mx

Davide Taibi
Università degli Studi dell'Insubria, Italy
Davide.taibi@uninsubria.it

C. Talhi
Concordia University, QC, Canada
talhi@ciise.concordia.ca

Davide Tosi
Università degli Studi dell'Insubria, Italy
davide.tosi@uninsubria.it

Emil Vassev
University College Dublin, Ireland
emil-vassev@lero.ie

L. Wang
Concordia University, QC,
Canada
wang@ciise.concordia.ca

Yuhong Yan
Concordia University, QC,
Canada
yuhong@cse.concordia.ca

Ludeng Zhao
Concordia University, QC,
Canada
ludeng.zhao@
    encs.concordia.ca

Towards Autonomic Specification of Distributed MARF with ASSL: Self-healing

Emil Vassev and Serguei A. Mokhov

Abstract. In this paper, we discuss our work towards self-healing property specification of an autonomic behavior in the Distributed Modular Audio Recognition Framework (DMARF) by using the Autonomic System Specification Language (ASSL). ASSL aids in enhancing DMARF with an autonomic middleware that enables it to perform in autonomous systems that theoretically require less-to-none human intervention. Here, we add an autonomic middleware layer to DMARF by specifying the core four stages of the DMARF's pattern-recognition pipeline as autonomic elements managed by a distinct autonomic manager. We devise the algorithms corresponding to this specification.

1 Introduction

The vision and metaphor of autonomic computing (AC) [7] is to apply the principles of self-regulation and complexity hiding. The AC paradigm emphasizes on reducing the workload needed to maintain complex systems by transforming them into self-managing autonomic systems. The idea is that software systems can manage themselves, and deal with on-the-fly occurring requirements automatically. This is the main reason why a great deal of research effort is devoted to the design and development of robust AC tools. Such a tool is the ASSL (Autonomic System Specification Language) framework, which helps AC researchers with problem formation,

Emil Vassev
Lero-the Irish Software Engineering Research Centre, University College Dublin,
Dublin, Ireland
e-mail: emil.vassev@lero.ie

Serguei A. Mokhov
Department of Computer Science and Software Engineering, Concordia University,
Montreal, QC, Canada
e-mail: mokhov@cse.concordia.ca

Roger Lee (Ed.): SERA 2010, SCI 296, pp. 1–15, 2010.
springerlink.com   © Springer-Verlag Berlin Heidelberg 2010

system design, system analysis and evaluation, and system implementation. In this work, we use ASSL [16, 15] to integrate autonomic features into the Distributed Modular Audio Recognition Framework (DMARF) – an intrinsically complex system composed of multi-level operational layers.

Problem Statement and Proposed Solution

Distributed MARF – DMARF – could not be used in autonomous systems of any kind "as-is" due to lack of provision for such a use by applications that necessitate autonomic requirements. Extending DMARF directly to support the said requirements is a major design and development effort for an open-source project.

In this work, we provide a proof-of-concept ASSL specification of one of the three core autonomic requirements for DMARF – self-healing (while the other two – self-protection and self-optimization are done in the work). In Appendix is the current outcome for the self-healing aspect and the rest of paper describes the methodology and related work behind it. Having the ASSL specification completed allows compiling it into the wrapper Java code as well as management Java code to provide an autonomic layer to DMARF to fulfill the autonomic requirement.

The rest of this paper is organized as follows. In Section 2, we review related work on AS specification and code generation. As a background to the remaining sections, Section 3 provides a brief description of both DMARF and ASSL frameworks. Section 4 presents the ASSL self-healing specification model for DMARF. Finally, Section 5 presents some concluding remarks and future work.

2   Related Work

IBM Research has developed a framework called Policy Management for Autonomic Computing (PMAC) [1], which provides a standard model for the definition of policies and an environment for the development of software objects that hold and evaluate policies. PMAC is used for the development and management of intelligent autonomic software agents. With PMAC, these agents have the ability to dynamically change their behavior, ability provided through a formal specification of policies encompassing the scope under which these policies are applicable. Moreover, policy specification includes the conditions under which a policy is in conformity (or has been violated), a set of resulting actions, goals or decisions that need to be taken and the ability to determine the relative value (priority) of multiple applicable actions, goals or decisions. [1]

3   Background

In this section, we introduce both the DMARF and the ASSL frameworks, thus needed to understand the specification models presented in Section 4.

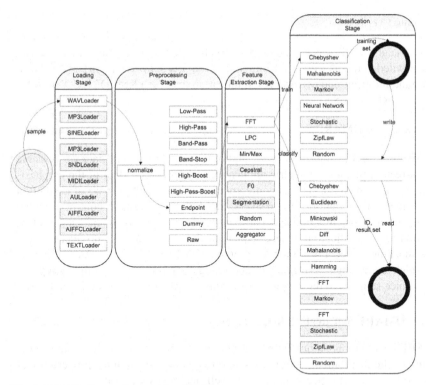

Fig. 1 MARF's Pattern Recognition Pipeline

3.1 Distributed MARF

DMARF [2, 5] is based on the classical MARF whose pipeline stages were made into distributed nodes.

Classical MARF

The Modular Audio Recognition Framework (MARF) [3] is an open-source research platform and a collection of pattern recognition, signal processing, and natural language processing (NLP) algorithms written in Java and put together into a modular and extensible framework. MARF can run distributively over the network, run stand-alone, or just act as a library in applications. The backbone of MARF consists of pipeline stages that communicate with each other to get the data they need in a chained manner. In general, MARF's pipeline of algorithm implementations is presented in Figure 1, where the implemented algorithms are in white boxes. The pipeline consists of four basic stages: sample loading, preprocessing, feature extraction, and training/classification. There are a number of applications that test MARF's functionality and serve as examples of how to use MARF's modules. One of the most prominent applications `SpeakerIdentApp` – Text-Independent Speaker Identification (who, gender, accent, spoken language, etc.).

Distributed Version

The classical MARF was extended [5, 4] to allow the stages of the pipeline to run as distributed nodes as well as a front-end, as roughly shown in Figure 2. The basic stages and the front-end perform communication over Java RMI [17], CORBA [8], and XML-RPC WebServices [9].

Applications of DMARF

High-volume processing of recorded audio, textual, or imagery data are possible pattern-recognition and biometric applications of DMARF. Most of the emphasis in this work is in audio, such as conference recordings with purpose of attribution of said material to identities of speakers. Similarly, processing a bulk of recorded phone conversations in a police department for forensic analysis and subject identification and classification, where sequential runs of the MARF instances on the same machine, especially a mobile equipment such as a laptop, PDA, cellphone, etc. which are not high-performance, an investigator has an ability of uploading collected voice samples to the servers constituting a DMARF-implementing network.

3.1.1 DMARF Self-healing Requirements

DMARF's capture as an autonomic system primarily covers the autonomic functioning of the distributed pattern-recognition pipeline. We examine properties that apply to DMARF and specify in detail the self-healing aspect of it.

If we look at the pipeline as a whole, we see that there should be at least one instance of the every stage somewhere on the network. There are four main core pipeline stages and application-specific stage that initiates pipeline processing. If one of the core stages goes offline completely, the pipeline stalls, so to recover one needs a replacement node, or recovery of the failed node, or to reroute the pipeline

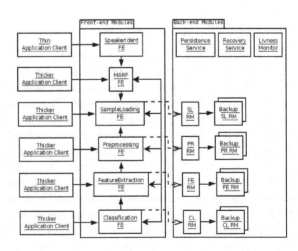

Fig. 2 The Distributed MARF Pipeline

through a different node with the same service functionality the failed one used to provide until the failed one recovers so a pipeline has to always self-heal and provide at least one pipeline route to be usable.

- A DMARF-based system should be able to recover itself in the form of replication to keep at least one route of the pipeline available. There are two types of replication – the replication of a service, which essentially means we increase the number of nodes per core stage (e.g. two different hosts provide preprocessing services as in active replication, so if one goes down, the pipeline is still not stalled; if both are up they can contribute to load balancing which is a part of the self-optimization aspect) and replication within the node itself. The latter replicas do not participate in the computation. They only receive updates and are on stand-by if the primary service goes down.
- If all nodes of a core stages go down, the stage preceding it is responsible to start up a temporary one on the host of the preceding stage, set it up to repair the pipeline. This is the hard replication needed to withstand stall faults, where it is more vulnerable and not fault-tolerant.
- In the second case, denoting passive replication of the same node (or even different nodes) losing a primary or a replica is not as serious as in the first case because such a loss does not produce a pipeline stall and it is easier to self-heal after a passive replica loss.
- Restart and recovery of the actual failed node without replicas is another possibility for self-healing for DMARF. Technically, it may be tried prior or after the replica kicks in.

3.2 ASSL

The Autonomic System Specification Language (ASSL) [15, 16] approaches the problem of formal specification and code generation of ASs within a framework. The core of this framework is a special formal notation and a toolset including tools that allow ASSL specifications be edited and validated. The current validation approach in ASSL is a form of consistency checking (handles syntax and consistency errors) performed against a set of semantic definitions. The latter form a theory that aids in the construction of correct AS specifications. Moreover, from any valid specification, ASSL can generate an operational Java application skeleton.

In general, ASSL considers autonomic system (ASs) as composed of autonomic elements (AEs) interacting over interaction protocols. To specify those, ASSL is defined through formalization tiers. Over these tiers, ASSL provides a multi-tier specification model that is designed to be scalable and exposes a judicious selection and configuration of infrastructure elements and mechanisms needed by an AS. The ASSL tiers and their sub-tiers (cf. Figure 3) are abstractions of different aspects of the AS under consideration. They aid not only to specifying the system at different levels of abstraction, but also to reducing the complexity, and thus, to improving the overall perception of the system.

There are three major tiers (three major abstraction perspectives), each composed of sub-tiers (cf. Figure 3):

- *AS tier* – forms a general and global AS perspective, where we define the general system rules in terms of *service-level objectives (SLO)* and *self-management policies*, *architecture topology*, and *global actions*, *events*, and *metrics* applied in these rules. Note that ASSL expresses policies with *fluents* (special states) [15, 16].
- *AS Interaction Protocol (ASIP) tier* – forms a communication protocol perspective, where we define the means of communication between AEs. The ASIP tier is composed of *channels*, *communication functions*, and *messages*.
- *AE tier* – forms a unit-level perspective, where we define interacting sets of individual AEs with their own behavior. This tier is composed of AE rules (*SLO* and *self-management policies*), an *AE interaction protocol* (AEIP), *AE friends* (a list of AEs forming a circle of trust), *recovery protocols*, special *behavior models* and *outcomes*, *AE actions*, *AE events*, and *AE metrics*.

The following elements describe some of the important sub-tiers in the ASSL specification model (cf. Figure 3):

- *AS/AE SLO* – Service-level objectives (SLO) are a higher-level form of behavioral specification that establishes objectives (e.g., performance). The ASSL concept assumes (but not restricts) that the AS service-level objectives (AS SLO) constitute a global task, the realization of which is to be decomposed into low-level AE service-level objectives (AE SLO).

```
I. Autonomic System (AS)
   * AS Service-level Objectives
   * AS Self-management Policies
   * AS Architecture
   * AS Actions
   * AS Events
   * AS Metrics
II. AS Interaction Protocol (ASIP)
   * AS Messages
   * AS Communication Channels
   * AS Communication Functions
III. Autonomic Element (AE)
   * AE Service-level Objectives
   * AE Self-management Policies
   * AE Friends
   * AE Interaction Protocol (AEIP)
     - AE Messages
     - AE Communication Channels
     - AE Communication Functions
     - AE Managed Resource Interface
   * AE Recovery Protocol
   * AE Behavior Models
   * AE Outcomes
   * AE Actions
   * AE Events
   * AE Metrics
```

Fig. 3 ASSL Multi-tier Specification Model

- *AS/AE Self-Management Policies* – The ASSL formal model specifies the four AC self-management policies (the so-called self-CHOP properties) [7]: *self-configuring*, *self-healing*, *self-optimizing*, and *self-protecting*.
- *AEIP* – The ASSL framework implies two communication protocols: *public* (ASIP) and *private* (AEIP). The autonomic element interaction protocol (AEIP) uses the same specification constructs as the ASIP does. Two AEs exchange messages over their AEIP only if they have an "agreement" on that.
- *Managed Resource Interface* – An AE typically controls one or more *managed elements*. In ASSL, this control is going over a special interface called *managed resource interface*. By having this interface specified, ASSL forms a simple communication model for interacting with the *managed elements*. This model forms an extra layer at the AEIP.
- *AS/AE Actions* – Actions are routines that can be performed in response to a *self-management policy*, *SLO*, or a *behavior model*.
- *AS/AE Events* – Events are one of the most important aspects in ASSL. They are used to specify many of the ASSL structures, such as *fluents*, *self-management policies*, *actions*, etc.
- *AS/AE Metrics* – Metrics constitute a set of parameters and observables, which the AEs can control.
- *AS Architecture* – At this sub-tier, the ASSL framework helps us to specify the topology of the AS. The architecture is specified as a correlation between the AEs or groups of AEs.
- *Behavior* – The ASSL framework specifies behavioral models by mapping *conditions* into potential *actions* and potential actions into *outcomes*.
- *Outcomes* – Outcomes are an abstraction for *action results*.

For further discussion on the ASSL multi-tier specification model and the ASSL framework toolset, please refer to [16, 13]. As part of the framework validation and in the course of a new currently ongoing research project at Lero (the Irish Software Engineering Research Center), ASSL has been used to specify autonomic properties and generate prototyping models of the NASA ANTS concept mission [12, 11] and NASA's Voyager mission [10].

4 Self-healing Model for DMARF

In order to make DMARF autonomic, we need to add *automicity* (autonomic computing behavior) to the DMARF behavior. Here we add a special *autonomic manager* (AM) to each DMARF stage. This makes the latter AEs, those composing an autonomic DMARF (ADMARF) capable of self-management.

Self-healing is one of the self-management properties that must be addressed by ADMARF (cf. Section 3.1.1). In the course of this project, we use ASSL to specify the self-healing behavior of ADMARF by addressing specific cases related to *node replacement* (service replica) and *node recovery* (cf. Section 3.1.1). The following elements present the specified self-healing algorithm:

- ADMARF monitors its run-time performance and in case of performance degradation notifies the problematic DMARF stages to start self-healing.
- Every notified DMARF stage (note that this is an AE) analyzes the problem locally to determine its nature: a node is down or a node is not healthy (does not perform well).
- If a node is down, the following node-replacement algorithm is followed by the AM of the stage:

  - AM strives to find a replica note of the failed one;
  - If replica found, next redirect computation to it;
  - If replica not found, report the problem. Note that the algorithm could be extended with a few more steps where the AM contacts the AM of the previous stage to organize pipeline reparation.

- If a node does not perform well, the following node-recovery algorithm is followed:

  - AM starts the recovery protocol for the problematic node;
  - If recovery successful do nothing;
  - If recovery unsuccessful, AM strives to find a replica node of the failed one;
  - If replica found, next redirect computation to it;
  - If replica not found, report the problem.

The following sections describe the ASSL specification of the self-healing algorithm revealed here. We specified this algorithm as an ASSL self-healing policy spread on both system (AS tier) and autonomic element (AE tier) levels where *events*, *actions*, *metrics*, and special *managed element interface functions* are used to incorporate the self-healing behavior in ADMARF (cf. Appendix). Note that due to space limitations Appendix presents a partial ASSL specification where only one AE (DMARF stage) is specified. The full specification specifies all the four DMARF stages.

4.1   AS Tier Specification

At the AS tier we specify the global ADMARF self-healing behavior. To specify the latter, we used an ASSL SELF_HEALING self-management policy structure (cf. Figure 4). Here we specified a single *fluent* mapped to an *action* via a *mapping*.

Thus, the inLowPerformance fluent is initiated by a lowPerformance Detected event and terminated by one of the events such as performance Normalized or performanceNormFailed. Here the inLowPerformance event is activated when special AS-level performance service-level objectives (SLO) get *degraded* (cf. Figure 5). Note that in ASSL, SLO are evaluated as *Booleans* based on their *satisfaction* and thus, they can be evaluated as *degraded* or *normal* [16]. Therefore, in our specification model, the lowPerformance Detected event will be activated anytime when the ADMARF performance goes down. Alternatively, the performanceNormalized event activates when the same performance goes up.

As it is specified, the AS-level `performance` SLO are a global task whose realization is distributed among the AEs (DMARF stages). Thus, the AS-level performance degrades if the performance of one of the DMARF stages goes down (cf. the FOREACH loop in Figure 5), thus triggering the SELF_HEALING policy.

In addition, the `performanceNormFailed` activates if an a special event (`selfHealingFailed`) occurs in the system. This event is specified at the AE tier (cf. Section 4.2) and reports that the local AE-level self-healing has failed. Although not presented in this specification, the `performanceNormFailed` event should be activated by any of the `performanceNormFailed` events specified for each AE (a DMARF stage). Moreover, once the `inLowPerformance` fluent gets initiated, the corresponding `startSelfHealing` action is executed (cf. Figure 5). This action simply triggers an AE-level event if the performance of that AE is degraded. The AE-level event prompts the SELF_HEALING policy at the AE level.

Fig. 4 AS Tier
SELF_HEALING Policy

```
ASSELF_MANAGEMENT {
    SELF_HEALING {
        // a performance problem has been detected
        FLUENT inLowPerformance {
            INITIATED_BY { EVENTS.lowPerformanceDetected }
            TERMINATED_BY { EVENTS.performanceNormalized,
                            EVENTS.performanceNormFailed }
        }
        MAPPING {
            CONDITIONS { inLowPerformance }
            DO_ACTIONS { ACTIONS.startSelfHealing }
        }
    }
} // ASSELF_MANAGEMENT
```

```
ASSLO {
    SLO performance {
        FOREACH member in AES {
            member.AESLO.performance
        }
    }
}
```

```
. . . . .
```

```
EVENTS { // events used in the fluents specificatio
    EVENT lowPerformanceDetected {
        ACTIVATION { DEGRADED { ASSLO.performance} } }
    EVENT performanceNormalized {
        ACTIVATION { NORMALIZED { ASSLO.performance} }
    EVENT performanceNormFailed {
        ACTIVATION { OCCURRED {
            AES.STAGE_AE.EVENTS.selfHealingFailed } }
    }
} // EVENTS
```

Fig. 5 AS Tier SLO and
Events

4.2   AE Tier Specification

At this tier we specify the self-healing policy for each AE in the ADMARF AS. Recall that the ADMARF's AEs are the DMARF stages enriched with a special autonomic manager each.

Appendix, presents the self-healing specification of one AE called STAGE_AE. Note that the latter can be considered as a generic AE and the specifications of the four AEs (one per DMARF stage) can be derived from this one.

Similar to the AS-level specification (cf. Section 4.1), here we specify a (but AE-level) SELF_HEALING policy with a set of *fluents* initiated and terminated by *events* and *actions* mapped to those *fluents* (cf. Appendix). Thus we specified three distinct fluents: inActiveSelfHealing, inFailedNodesDetected, inProblematicNodesDetected, each mapped to an AE-level action. The first fluent gets initiated when a mustDoSelfHealing event occurs in the system. That event is triggered by the AS-level startSelfHealing action in the case the performance SLO of the AE get degraded (cf. Appendix).

Here the performance SLO of the AE are specified as a Boolean expression over two ASSL metrics, such as the numberOfFailedNodes metric and the equivalent numberOfProblematicNodes (cf. Figure 6). Whereas the former measures the number of *failed notes* in the DMARF stage, the latter measures the number of *problematic nodes* in that stage. Both metrics are specified as RESOURCE metrics, i.e., observing a managed resource controlled by the AE [16]. Note that the managed resource here is the DMARF stage itself. Thus, as those metrics are specified (cf. Appendix) they get updated by the DMARF stage via special *interface functions* embedded in the specification of a STAGE_ME managed element (cf. Appendix). In addition, both metrics are set to accept only a zero value (cf. Figure 7), thus set in the so-called metric THRESHOLD_CLASS [16]. The latter determines rules for *valid* and *invalid* metric values. Since in ASSL metrics are evaluated as Booleans (valid or invalid) based on the value they are currently holding, the performance SLO (cf. Figure 6) will get degraded if one of the two defined metrics, the numberOfFailedNodes or numberOfProblematicNodes becomes *invalid*, i.e., if the DMARF stage reports that there is one or more *failed* or *problematic* nodes.

The inActiveSelfHealing fluent prompts the analyzeProblem action execution (cf. Appendix). The latter uses the STAGE_ME managed element's interface functions to determine the nature of the problem – is it a node that failed or it is a node that does not perform well. Based on this, the action triggers a

```
AESLO {
  SLO performance {
    METRICS.numberOfFailedNodes
    AND
    METRICS.numberOfProblematicNodes
  }
}
```

Fig. 6 AE Tier SLO

Fig. 7 AE Tier Metric
Threshold Class

```
VALUE { 0 }
THRESHOLD_CLASS { Integer [0] }
```

Fig. 8 AE Tier STAGE_ME
Functions

```
// runs the replica of a failed nodee
INTERFACE_FUNCTION runNodeReplica {
    PARAMETERS { DMARFNode node }
    ONERR_TRIGGERS { EVENTS.nodeReplicaFailed }
}
// recovers a problematic node
INTERFACE_FUNCTION recoverNode {
    PARAMETERS { DMARFNode node }
    ONERR_TRIGGERS { EVENTS.nodeCannotBeFixed }
}
```

Fig. 9 Event
mustSwitchToNodeReplica

```
EVENT mustSwitchToNodeReplica {
    ACTIVATION {
        OCCURRED { EVENTS.nodeCannotBeFixed }
    }
}
```

mustSwitchToNodeReplica event or a mustFixNode event respectively. Each one of those events initiates a fluent in the AE SELF_HEALING policy to handle the performance problem. Thus, the inFailedNodesDetected fluent handles the case when a node has failed and its replica must be started and the inProblematicNodesDetected fluent handles the case when a node must be recovered. Here the first fluent prompts the execution of the startReplicaNode action and the second prompts the execution of the fixProblematicNode action. Internally, both actions call interface functions of the STAGE_ME managed element. Note that those functions trigger erroneous events if they do not succeed (cf. Figure 8). Those events terminate fluents of the AE SELF_HEALING policy (cf. Appendix). It is important to mention that the inFailedNodesDetected fluent gets initiated when the mustSwitchToNodeReplica event occurs in the system. The latter is triggered by the analyzeProblem action. Furthermore, according to its specification (cf. Figure 9) the same event activates if a nodeCannotBeFixed event occurs in the system, which is due to the inability of the recoverNode interface function to recover the problematic node (cf. Figure 8). Therefore, if a node cannot be recovered the inFailedNodesDetected fluent will be initiated in an attempt to start the replica of that node. Note that this conforms to the self-healing algorithm presented in Section 4.

5   Conclusion and Future Work

We constructed a self-healing specification model for ADMARF. To do so we devised algorithms with ASSL for the pipelined stages of the DMARF's pattern recognition pipeline. When fully-implemented, the ADMARF system will be able to fully function in autonomous environments, be those on the Internet, large multimedia processing farms, robotic spacecraft that do their own analysis, or simply even pattern-recognition research groups that can rely more on the availability of their systems that run for multiple days, unattended. Although far from a full specification model for ADMARF, we have attempted to provide didactic evidence of how ASSL can help us achieve desired automicity in DMARF along with our related work on the self-protecting and self-optimizing properties that are documented elsewhere [6, 14].

Future Work

Some work on both projects, DMARF and ASSL is still on-going, that, when complete, will allow a more complete realization of ADMARF. Some items of the future work are as follows:

- We plan producing a specification of the self-configuration aspect of ADMARF.
- We plan on releasing the Autonomic Specification of DMARF, ADMARF as open-source.

Acknowledgements. This work was supported in part by an IRCSET postdoctoral fellowship grant (now termed as EMPOWER) at University College Dublin, Ireland, by the Science Foundation Ireland grant 03/CE2/I303_1 to Lero – the Irish Software Engineering Research Centre, and by the Faculty of Engineering and Computer Science of Concordia University, Montreal, Canada.

Appendix

```
// ASSL self-healing specification model for DMARF

AS DMARF {
  TYPES { DMARFNode }
  ASSLO {
    SLO performance {
      FOREACH member in AES { member.AESLO.performance }
    }
  }

  ASSELF_MANAGEMENT {
    SELF_HEALING {
      // a performance problem has been detected
      FLUENT inLowPerformance {
        INITIATED_BY { EVENTS.lowPerformanceDetected }
        TERMINATED_BY {
          EVENTS.performanceNormalized, EVENTS.performanceNormFailed }
      }
      MAPPING {
        CONDITIONS { inLowPerformance }
        DO_ACTIONS { ACTIONS.startSelfHealing }
      }
    }
  } // ASSELF_MANAGEMENT

  ACTIONS {
    ACTION IMPL startSelfHealing {
      GUARDS { ASSELF_MANAGEMENT.SELF_HEALING.inLowPerformance }
      TRIGGERS {
        IF NOT AES.STAGE_AE.AESLO.performance THEN
          AES.STAGE_AE.EVENTS.mustDoSelfHealing
        END
      }
    }
  } // ACTIONS

  EVENTS { // these events are used in the fluents specification
    EVENT lowPerformanceDetected {
      ACTIVATION { DEGRADED { ASSLO.performance} } }
    EVENT performanceNormalized {
      ACTIVATION { NORMALIZED { ASSLO.performance} } }
```

```
      EVENT performanceNormFailed {
        ACTIVATION { OCCURRED { AES.STAGE_AE.EVENTS.selfHealingFailed } } }
    } // EVENTS
} // AS DMARF

AES {
  AE STAGE_AE {
    VARS { DMARFNode nodeToRecover }
    AESLO {
      SLO performance {
        METRICS.numberOfFailedNodes AND METRICS.numberOfProblematicNodes
      }
    }
    AESELF_MANAGEMENT {
      SELF_HEALING {
        FLUENT inActiveSelfHealing {
          INITIATED_BY { EVENTS.mustDoSelfHealing }
          TERMINATED_BY { EVENTS.selfHealingSuccessful, EVENTS.selfHealingFailed }
        }
        FLUENT inFailedNodesDetected {
          INITIATED_BY { EVENTS.mustSwitchToNodeReplica }
          TERMINATED_BY { EVENTS.nodeReplicaStarted, EVENTS.nodeReplicaFailed }
        }
        FLUENT inProblematicNodesDetected {
          INITIATED_BY { EVENTS.mustFixNode }
          TERMINATED_BY { EVENTS.nodeFixed, EVENTS.nodeCannotBeFixed }
        }
        MAPPING {
          CONDITIONS { inActiveSelfHealing }
          DO_ACTIONS { ACTIONS.analyzeProblem }
        }
        MAPPING {
          CONDITIONS { inFailedNodesDetected }
          DO_ACTIONS { ACTIONS.startReplicaNode }
        }
        MAPPING {
          CONDITIONS { inProblematicNodesDetected }
          DO_ACTIONS { ACTIONS.fixProblematicNode }
        }
      }
    } // AESELF_MANAGEMENT
    AEIP {
      MANAGED_ELEMENTS {
        MANAGED_ELEMENT STAGE_ME {
          INTERFACE_FUNCTION countFailedNodes { RETURNS { Integer  } }
          INTERFACE_FUNCTION countProblematicNodes { RETURNS { Integer } }
          // returns the next failed node
          INTERFACE_FUNCTION getFailedNode { RETURNS { DMARFNode } }
          // returns the next problematic node
          INTERFACE_FUNCTION getProblematicNode { RETURNS { DMARFNode } }
          // runs the replica of a failed nodee
          INTERFACE_FUNCTION runNodeReplica {
            PARAMETERS { DMARFNode node }
            ONERR_TRIGGERS { EVENTS.nodeReplicaFailed }
          }
          // recovers a problematic node
          INTERFACE_FUNCTION recoverNode {
            PARAMETERS { DMARFNode  node }
            ONERR_TRIGGERS { EVENTS.nodeCannotBeFixed }
          }
        }
      }
    } // AEIP
    ACTIONS {
      ACTION analyzeProblem {
        GUARDS { AESELF_MANAGEMENT.SELF_HEALING.inActiveSelfHealing }
        VARS { BOOLEAN failed }
        DOES {
          IF METRICS.numberOfFailedNodes THEN
            AES.STAGE_AE.nodeToRecover = call
              AEIP.MANAGED_ELEMENTS.STAGE_ME.getFailedNode;
            failed = TRUE
          END
          ELSE
            AES.STAGE_AE.nodeToRecover = call
              AEIP.MANAGED_ELEMENTS.STAGE_ME.getProblematicNode;
            failed = FALSE
          END
        }
        TRIGGERS {
          IF failed THEN EVENTS.mustSwitchToNodeReplica END
          ELSE EVENTS.mustFixNode END
        }
        ONERR_TRIGGERS { EVENTS.selfHealingFailed }
      }
      ACTION startReplicaNode {
        GUARDS { AESELF_MANAGEMENT.SELF_HEALING.inFailedNodesDetected }
        DOES {
          call AEIP.MANAGED_ELEMENTS.STAGE_ME.runNodeReplica(AES.STAGE_AE.nodeToRecover) }
```

```
      TRIGGERS { EVENTS.nodeReplicaStarted }
    }
    ACTION fixProblematicNode {
      GUARDS { AESELF_MANAGEMENT.SELF_HEALING.inProblematicNodesDetected }
      DOES {
        call AEIP.MANAGED_ELEMENTS.STAGE_ME.recoverNode(AES.STAGE_AE.nodeToRecover) }
      TRIGGERS { EVENTS.nodeFixed}
    }
  } // ACTIONS
  EVENTS {
    EVENT mustDoSelfHealing { }
    EVENT selfHealingSuccessful {
      ACTIVATION {
        OCCURRED { EVENTS.nodeReplicaStarted }
        OR
        OCCURRED { EVENTS.nodeFixed }
      }
    }
    EVENT selfHealingFailed {
      ACTIVATION { OCCURRED {  EVENTS.nodeReplicaFailed } }
    }
    EVENT mustSwitchToNodeReplica {
      ACTIVATION { OCCURRED {  EVENTS.nodeCannotBeFixed } }
    }
    EVENT nodeReplicaStarted { }
    EVENT nodeReplicaFailed {  }
    EVENT mustFixNode { }
    EVENT nodeFixed {  }
    EVENT nodeCannotBeFixed {  }
  } // EVENTS
  METRICS {
    // increments when a failed node has been discovered
    METRIC numberOfFailedNodes {
      METRIC_TYPE { RESOURCE  }
      METRIC_SOURCE { AEIP.MANAGED_ELEMENTS.STAGE_ME.countFailedNodes }
      DESCRIPTION {"counts failed nodes in the MARF stage"}
      VALUE { 0 }
      THRESHOLD_CLASS {  Integer [0] } // valid only when holding 0 value
    }
    // increments when a problematic node has been discovered
    METRIC numberOfProblematicNodes {
      METRIC_TYPE { RESOURCE  }
      METRIC_SOURCE { AEIP.MANAGED_ELEMENTS.STAGE_ME.countProblematicNodes }
      DESCRIPTION {"counts nodes with problems in the MARF stage"}
      VALUE { 0 }
      THRESHOLD_CLASS {  Integer [0] } // valid only when holding 0 value
    }
  }
 }
}
```

References

1. IBM Tivoli: Autonomic computing policy language. Tech. rep., IBM Corporation (2005)
2. Mokhov, S.A.: On design and implementation of distributed modular audio recognition framework: Requirements and specification design document. Project report (2006), http://arxiv.org/abs/0905.2459 (last viewed May 2009)
3. Mokhov, S.A.: Choosing best algorithm combinations for speech processing tasks in machine learning using MARF. In: Bergler, S. (ed.) Canadian AI 2008. LNCS (LNAI), vol. 5032, pp. 216–221. Springer, Heidelberg (2008)
4. Mokhov, S.A., Huynh, L.W., Li, J.: Managing distributed MARF's nodes with SNMP. In: Proceedings of PDPTA 2008, vol. II, pp. 948–954. CSREA Press, Las Vegas (2008)
5. Mokhov, S.A., Jayakumar, R.: Distributed modular audio recognition framework (DMARF) and its applications over web services. In: Proceedings of TeNe 2008. Springer, University of Bridgeport (2008)
6. Mokhov, S.A., Vassev, E.: Autonomic specification of self-protection for Distributed MARF with ASSL. In: Proceedings of C3S2E 2009, pp. 175–183. ACM, New York (2009)
7. Murch, R.: Autonomic Computing: On Demand Series. IBM Press, Prentice Hall (2004)
8. Sun: Java IDL. Sun Microsystems, Inc. (2004), http://java.sun.com/j2se/1.5.0/docs/guide/idl/index.html

9. Sun Microsystems: The java web services tutorial (for Java Web Services Developer's Pack, v2.0). Sun Microsystems, Inc. (2006),
   http://java.sun.com/webservices/docs/2.0/tutorial/doc/
10. Vassev, E., Hinchey, M.: ASSL specification model for the image-processing behavior in the NASA Voyager mission. Tech. rep., Lero - The Irish Software Engineering Research Center (2009)
11. Vassev, E., Hinchey, M., Paquet, J.: A self-scheduling model for NASA swarm-based exploration missions using ASSL. In: Proceedings of the Fifth IEEE International Workshop on Engineering of Autonomic and Autonomous Systems (EASe 2008), pp. 54–64. IEEE Computer Society, Los Alamitos (2008)
12. Vassev, E., Hinchey, M.G., Paquet, J.: Towards an ASSL specification model for NASA swarm-based exploration missions. In: Proceedings of the 23rd Annual ACM Symposium on Applied Computing (SAC 2008) - AC Track, pp. 1652–1657. ACM, New York (2008)
13. Vassev, E., Mokhov, S.A.: An ASSL-generated architecture for autonomic systems. In: Proceedings of C3S2E 2009, pp. 121–126. ACM, New York (2009)
14. Vassev, E., Mokhov, S.A.: Self-optimization property in autonomic specification of Distributed MARF with ASSL. In: Shishkov, B., Cordeiro, J., Ranchordas, A. (eds.) Proceedings of ICSOFT 2009, vol. 1, pp. 331–335. INSTICC Press, Sofia (2009)
15. Vassev, E., Paquet, J.: ASSL – Autonomic System Specification Language. In: Proceedings if the 31st Annual IEEE/NASA Software Engineering Workshop (SEW-31), pp. 300–309. NASA/IEEE. IEEE Computer Society, Baltimore (2007)
16. Vassev, E.I.: Towards a framework for specification and code generation of autonomic systems. Ph.D. thesis, Department of Computer Science and Software Engineering, Concordia University, Montreal, Canada (2008)
17. Wollrath, A., Waldo, J.: Java RMI tutorial. Sun Microsystems, Inc. (1995–2005),
   http://java.sun.com/docs/books/tutorial/rmi/index.html

Repairing Service Compositions in a Changing World

Yuhong Yan, Pascal Poizat, and Ludeng Zhao

1 Introduction and Motivation

Service-Oriented Computing (SOC) promotes the reuse of loosely coupled and distributed entities, namely services, and their automatic composition into value-added applications. An issue in SOC is to fulfill this promise with the development of models and algorithms supporting composition in an automatic (and automated) way, generating business processes from a set of available services and composition requirements, *e.g.,* descriptions of end-user needs or business goals. Within the context of the widely accepted service-oriented architecture, *i.e.,* Web services, this issue is currently known as the *Web Service Composition (WSC)* problem. WSC has been widely addressed in the past few years [15]. *AI planning* is increasingly applied to WSC due to its support for automatic composition from under-specified requirements [19]. An AI planning based algorithm usually provides a one-time solution, *i.e.,* a *plan*, for a composition request. However, in the real and open world, change occurs frequently. Services may appear and disappear at any time in a unpredictable way, *e.g.,* due to failure or the user mobility when services get out of reach. Service disappearance requires changing the original plan, and may cause some goals to be unreachable. End-user needs or business goals may also change over time. For example, one may want to add sightseeing functionality to a composition when arrived at a trip destination.

Yuhong Yan
Concordia University, Montreal, Canada
e-mail: yuhong@cse.concordia.ca

Pascal Poizat
University of Evry Val d'Essonne, Evry, and LRI UMR 8623 CNRS, Orsay, France
e-mail: pascal.poizat@lri.fr

Ludeng Zhao
Concordia University, Montreal, Canada
e-mail: ludeng.zhao@encs.concordia.ca

Roger Lee (Ed.): SERA 2010, SCI 296, pp. 17–36, 2010.

The challenge we address in this paper is to define techniques supporting adaptation for a plan as a reaction to changes (available services or composition requirements). Basically, this may be achieved in two distinct ways. The first one is to perform a comprehensive *replanning* from the current execution state of a plan. Another way is trying to *repair* the plan in response to changes, reusing most of the original plan when possible. We believe plan repair is a valuable solution to WSC in a changing world because:

- it makes it possible to retain the effects of a partially executed plan, which is better than throwing it away and having to roll back its effects;
- even if, in theory, modifying an existing plan is no more efficient than a comprehensive replanning in the worst case [23], we expect that plan repair is, in practice, often more efficient than replanning since a large part of a plan is usually still valid after a change has occured;
- commitment to the unexecuted services can be kept as much as possible, which can be mandatory for business/security reasons;
- finally, the end-user may prefer to use a resembling plan resulting from repair than a very different plan resulting from replanning.

Our proposal is based on planning graphs [13]. They enable a compact representation of relations between services and model the whole problem world. Even with some changes, part of a planning graph is still valid. Therefore, we think that the use of planning graphs can be better than other techniques in solving plan adaptation problems. We first identify the new composition problem with the new set of available services and new goals. The disappeared services are removed from the original planning graph and new goals are added to the goal level, yielding a partial planning graph. The repair algorithm "regrows" this partial planning graph wherever the heuristic function tells the unimplemented goals and broken preconditions can be satisfied. Its objective is not to build a full planning graph, but to fast search for a feasible solution, while maximally reuse whatever is in the partial graph. Compared to replanning, the repair algorithm only constructs a part of a full planning graph until a solution is found. It can be faster than replanning, especially when the composition problem does not change too much and a large part of the original planning graph is still valid. In our experiments, we have proved this. Our experiments also show that the solutions from repair have the same quality as those from replanning. However, in some cases, our repair algorithm may not find existing solutions where replanning would, which is the trade off of its speed.

As far as Web services are concerned, we take into consideration their WSDL interfaces extended with semantic information for inputs and outputs of operations. We suppose services are stateless, therefore we do not consider the internal behavior of Web services. Accordingly, composition requirements are data-oriented and not based on some required conversation. These choices are consistent with many approaches for WSC, *e.g.*, [17] [18] [12]. More importantly, it suits to the Web Service Challenge [4] which enables us to evaluate our proposal on big-size data sets.

The remaining of the paper is as follows. In section 2 we present our formal models for services (including semantic information), composition requirements, and Web service composition. We also discuss the relations between these models and real service languages. Finally, we introduce graph planning and its application to WSC, and present its use on a simple example. In sections 3 and 4 we respectively address our model for change in a service composition and introduce the evaluation criteria for replanning and repair. Our repair algorithm is presented in section 5 and is evaluated in section 6. We end up with discussion on related work, conclusions and perspectives.

2 Formal Modelling

2.1 Semantic Models

In order to enable automatic service discovery and composition from user needs, some form of semantics has to be associated to services. Basically, the names of the services' provided operations could be used, but one can hardly imagine and, further, achieve interoperability at this level in a context where services are designed by different third parties. Adaptation approaches [21] [6] have proposed to rely on so-called adaptation contracts that must be given (totally or partly) manually. To avoid putting this burden on the user, semantics based on shared accepted ontologies may be used instead to provide fully-automatic compositions. Services may indeed convey two kinds of semantic information. The first one is related to data that is transmitted along with messages. Let us suppose a service with an operation providing hotel rooms. It can be semantically described as taking as input semantic information about **travelcity, fromdate, todate** and **username** and providing an **hotel-registration**. The second way to associate semantic information to services is related to functionalities, or capacities, that are fulfilled by services. Provided each service has a single capacity, this can be treated as a specific output in our algorithms, *e.g.*, the abovementioned service could be described with an additional semantic output, **hotelbookingcapacity**, or we can even suppose its capacity is self-contained in its outputs (here, **hotelregistration**).

In our approach, semantic information is supported with a structure called *Data Semantic Structure* (Def. 1).

Definition 1. *A Data Semantic Structure (DSS) $\mathscr{D}$ is a couple (D, R^D) where D is a set of concepts that represent the semantics of some data, and $R^D = \{R_i^D : 2^D \to 2^D\}$ is a set of relations between concepts.*

The members of R^D define how, given some data, other data can be produced. Given two sets, $D_1, D_2 \subseteq D$, we say that D_2 can be obtained from D_1, and we write $D_1 \rightsquigarrow_{R_i^D} D_2$, or simply $D_1 \rightsquigarrow D_2$, when $\exists R_i^D \in R^D, R_i^D(D_1) = D_2$. Some instances of the R_i^D relations – representing composition, decomposition and casting – together with implementation rules are presented in [3]. Here we propose a generalization of these.

2.2 Service and Service Composition Models

A Web service can be considered for composition as a function that takes input(s) and returns output(s) (Def. 2). We abstract here from the way how input and output semantic data, called parameters in the sequel, relate to service operations. An example of this where relations are defined between WSDL message types and service model parameters through the use of XPath can be found in [16].

Definition 2. *Being given a DSS* (D, R^D), *a Web service w is a tuple* (in, out) *where* $in \subseteq D$ *denote the input parameters of w and* $out \subseteq D$ *denote the output parameters of w. We denote* $in^- \subseteq in$ *the set of input parameters that are consumed by the service.*

DSS support data transformation using relations in R^D. This enables service composition where mismatch would usually prevent it (see Section 2, 2.5 for examples of these). For the sake of uniformity, being given a DSS (D, R^D), for each R_i^D in R^D, for each $D_1 \leadsto_{R_i^D} D_2$, we define a data adaptation service $w_{R_i} = (D_1, D_2)$. Such a service can be automatically implemented, either as a service or as a reusable standalone piece of code in any implementation language supporting XPath and assignment such as WS-BPEL.

If the output parameters of a set of services can produce at least one of the input parameters of another service, we say they can be connected (Def. 3).

Definition 3. *Assume every parameter in the parameter set* $D_W = \{d_1, d_2, \ldots, d_k\}$ *is an output parameter of one of the Web services in a set* $W = \{w_1, w_2, \ldots, w_m\}$, *i.e.,* $W = \{w_i \mid \exists d_j \in D_W, d_j \in out(w_i), i = 1, \ldots, m\}$. *If* $\{d_1, d_2, \ldots, d_k\} \leadsto \{d_n\}$, *and* $d_n \in in(w_n)$, *every service in W can be connected to* w_n, *annotated as* $w_i \triangleright w_n$.

Finally, the last input for any WSC algorithm is the description of the composition requirements corresponding to the user needs (Def. 4).

Definition 4. *Being given a DSS* (D, R^D), *a composition requirement is a couple* (D_U^{in}, D_U^{out}) *with* $D_U^{in} \subseteq D$ *is the set of provided (or input) parameters and* $D_U^{out} \subseteq D$ *is the set of required (or output) parameters.*

The objective of a WSC algorithm may now be formally described. Given a set of available Web services, a structure (DSS) describing the semantic information that is associated to the services, and a composition requirement, service composition is to generate a connected subset of the services that satisfies the composition requirement (Def. 5).

Definition 5. *A composition requirement* (D_U^{in}, D_U^{out}) *is satisfied by a set of connected Web services* $W = \{w_1, \ldots, w_n\}$ *iff,* $\forall i \in \{1, \ldots, n\}$:

- $\forall d \in in(w_i), D_U^{in} \cup out(w_1) \cup \ldots \cup out(w_{i-1}) \leadsto d$ *and*
- $\forall d \in D_U^{out}, D_U^{in} \cup out(w_1) \cup \ldots \cup out(w_n) \leadsto d$

A composition requirement is satisfied iff there is at least one set of connected services satisfying it.

2.3 Models and Languages

The interface (operations and types of exchanged messages) of a Web service is described in a WSDL [26] file. DSSs are a subclass of what can be described using OWL [24], a very expressive language for describing super(sub)-classes, function domains and ranges, equivalence and transition relations, etc. We can extend OWL for other relations as well, *e.g.*, a parameter can correspond to the (de)composition of a set of parameters according to some functions. WSDL can be extended to reference OWL semantic annotations [4]. The SAWSDL standard [25] can also be used for this.

Some service input parameters are *informatic*, and can be therefore used without limitation. However, other service input parameters are *consumable*, meaning that they can be used only once. For example, given an order processing service, orders are consumable: once processed, orders cannot be used by other services. OWL's `rdf:property` can be extended to describe parameters consumability properties.

In this work we suppose services are stateless and do not feature a behavioural description specifying in which ordering operations are to be called. Hence, without loss of generality, we have supposed each service has only one operation. Whenever a service has more than one, we can use indexing, *e.g.*, service w with operations o_1 and o_2 becomes services $w.o_1$ and $w.o_2$.

2.4 Web Service Composition as Planning

AI planning [10] has been successfully applied to solve the WSC problem through its encoding as a planning problem [19, 15] (Def. 6). The following definitions in this subsection are modified from [10].

Definition 6. *A planning problem is a triple* $P = ((S,A,\gamma), s_0, g)$*, where*

- *S is a set of states, with a state s being a subset of a finite set of proposition symbols L, where $L = \{p_1,\dots,p_n\}$.*
- *A is a set of actions, with an action a being a triple $(precond, effects^-, effects^+)$ where $precond(a)$ denotes the preconditions of a, and $effects^-(a)$ and $effects^+(a)$, with $effects^-(a) \cap effect^+(a) = \emptyset$, denote respectively the negative and the positive effects of a.*
- *γ is a state transition function such that, for any state s where $precond(a) \subseteq s$, $\gamma(s,a) = (s - effects^-(a)) \cup effects^+(a)$.*
- *$s_0 \in S$ is the initial state.*
- *$g \subseteq L$ is a set of propositions called goal propositions (or simply goal).*

A plan is any sequence of actions $\pi = \langle a_1,\dots,a_k \rangle$, where $k \geq 0$.

The WSC problem can be mapped to a planning problem as follows:

- each service, w, is mapped to an action with the same name, w. The input parameters of the service, $in(w)$, are mapped to the action's preconditions, $precond(w)$, and its output parameters, $out(w)$, are mapped to the action's positive effects,

$effects^+(w)$. Consumable parameters, $in^-(w)$ are also mapped to negative effects, $effects^-(w)$.

- the input parameters of the composition requirement, D_U^{in}, are mapped to the initial state, s_0.
- the output parameters of the composition requirement, D_U^{out}, are mapped to the goal, g.

Different planning algorithms have been proposed to solve planning problems, *e.g.*, depending on whether they are building the graph structure underlying a planning problem in a forward (from initial state) or backward (from goal) manner. In our work we propose to use a planning graph [10] (Def.7) based algorithm.

Definition 7. *In a planning graph, a layered plan is a sequence of sets of actions* $\langle \pi_1, \pi_2, \ldots, \pi_n \rangle$, *in which each* π_i *(i = 1,...,n) is independent (see Def. 8).* π_1 *is applicable to* s_0. π_i *is applicable to* γ (s_{i-2}, π_{i-1}) *when* $i = 2, \ldots, n$. $g \subseteq \gamma(\ldots(\gamma(\gamma(s_0, \pi_1), \pi_2) \ldots \pi_n)$.

A planning graph iteratively expands itself one level at a time. The process of graph expansion continues until either it reaches a level where the proposition set contains all goal propositions or a fixed point level. The goal cannot be attained if the latter happens first. Definition 8 defines the independence of two actions. Two actions can also exclude each other due to the conflicts of their effects. We can add both independent and dependent actions in one action layer in the planning graph. However, two exclusive actions cannot appear in the same plan. The planning graph searches backward from the last level of the graph for a solution. It is known that a planning graph can be constructed in polynomial time.

Definition 8. *In a planning graph, two actions a and b are independent iff they satisfy* $effects^-(a) \cap [precond(b) \cup effects^+(b)] = \emptyset$, *and* $effects^-(b) \cap [precond(a) \cup effects^+(a)] = \emptyset$. *A set of actions is independent when its actions are pairwise independent.*

2.5 Example

In order to demonstrate our model and our planning algorithm, we give a simple travelling example. Evaluation on benchmark data with a much larger set of services is presented in section 6.

First we introduce the DSS for our example. The concepts we use correspond to:

- users, uinfo (made up of uname and ucity),
- departure and return dates, fromdate and todate,
- departure and return cities (resp. countries), depcity and depcountry (resp. destcity and destcountry),
- travelling alerts, travelalert,
- flight requests, flightreq, and finally,
- registration information for planes and hotels, planereg and hotelreg.

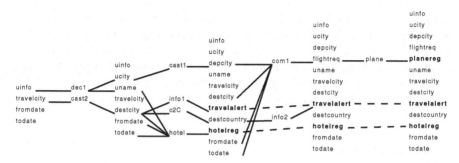

Fig. 1 Travel example - Planning graph

Additionally, different relations exist between these concepts:

- subsumption: from ucity to depcity and from travelcity to destcity,
- decomposition: from uinfo to uname and ucity
- composition: from depcity, destcity, fromdate, and todate, to flightreq.

As explained before, these four relations will be supported by data adaptation services (respectively cast1, cast2, dec1, and comp1) that can automatically be implemented. The repository of available services contains the following services:

- info1({destcity}, {travelalert}),
- info2({destcountry}, {travelalert}),
- c2C({destcity}, {destcountry}),
- plane({flightreq,uname}, {planereg}), and
- hotel({travelcity,fromdate,todate,uname}, {hotelreg}).

Finally, the user request is:
({uinfo,travelcity,fromdate,todate}, {planereg,hotelreg,travelalert}).

Applying our planning algorithm to this example, we get the planning graph in figure 1. Identity links (when some data is kept for the next data layer) are not represented for clarity, but for required parameters (dashed lines).

From this planning graph, backtracking from the required data (in bold), we can compute two solutions (plans), namely (dec1∥cast2);(cast1∥info1∥hotel);com1;plane and (dec1∥cast2);(cast1∥c2C∥hotel);(com1∥info2);plane, where ; represents the sequence of service calls and ∥ service calls in parallel (flow in BPEL). One may note that without the adaptation enabled by DSS relations (cast1, cast2, dec1, and comp1), no composition would be possible at all.

3 Adapt to the Changing World

The WSC problem should be considered in an open world. An open world is always changing. First, Web services appear and disappear all the time (***environment changes***). New services provide new opportunities to achieve some better effects than the original plan. Therefore, we may consider to use new Web services and replace some old ones. Second, during the execution of the process, business goals may change due to shifting business interests (***goal changes***). Hence, new goals may be added and old ones may be replaced. Third, repairing a failed process also generates goal changes, as well as plan changes (***fault caused changes***). We can diagnose the faulty Web services which are responsible for a halting process [27]. We need to eliminate the failure effects, which includes rolling back the executed Web services and find replacement of the faulty Web services.

Environment Changes

Services can be added and removed in the environment. We model it as a simple operation on the service set as in Equation 1.

$$W' = W - W_{disappear} + W_{appear} \tag{1}$$

Goal Changes

We assume the removed goals and the additional goals are given (probably manually generated by user). We just simply update the goals as in Equation 2. We assume that there is no conflict (mutually exclusive propositions) in g. It is possible to automatically generating new business goals. For example, based on user profile, we can suggest sport activities instead of visiting museum for a sport lover to fill a time gap due to meeting cancellation in a travel plan. Generating new goals is not studied in this paper.

$$g' = g - g_{remove} + g_{new\_business} \tag{2}$$

Fault Caused Changes

We can diagnose the faulty Web services as in [27]. The real cause of the exception (the fault) can occur before any symptom (the exception) is observed. When we observe any symptom, the faulty services are already executed. For an identified faulty Web service w, we have some options. Option 1: we can remove the faulty service w, as if the service had disappeared. Option 2: some exceptions are due to format error. This kind of errors can possibly be fixed by finding another service to do format conversion. In this case, we can leave the service unchanged. Option 3: we can identify that the error is only at a certain output parameter p. Then we can project the function of w on the rest of the rest of its outputs, *i.e.*, remove p from $out(w)$, as in Equation 3.

$$out'(w) = out(w) - p \tag{3}$$

When we remove an executed service, *e.g.*, an executed faulty service, it requires to roll back the effects of this service. If a service w needs to roll back, the goals should be updated as in Equation 4. We do not constraint how the effects of a service can be canceled. It can be another service, or a roll back operation of the same service.

$$g = g - effects^+(w) + effects^-(w) \tag{4}$$

We generally can solve the adaptation problem in two ways **replanning** or **repair**. By **replanning**, we mean that we try to solve a newly constructed planning problem $P' = ((S', A', \gamma'), s_0, g')$, where A' is an updated set of available Web services, g' is a updated set of goals, with S' and γ' changed accordingly, from scratch. By **repair**, we mean that we try to fix the existing, but broken, plans. We consider that planning graph is a good candidate for adaptation, because a planning graph models the whole problem world. If the problem changes, part of the planning graph is still valid. In this paper, we present a repair algorithm that can grow the partial valid planning graph until a solution is found. Below is an intuitive idea to show that repair can be faster. However, the evaluation will be done in a systematic way in the rest of the paper.

Example: The services in the following table are available. a is known, and e is the request.

Table 1 Available Services

$A2BC : a \rightarrow b, c$	$A2D : a \rightarrow d$	$C2E : c \rightarrow e$
$D2E : d \rightarrow e$	$D2F : d \rightarrow f$	$F2G : f \rightarrow g$
$F2H : f \rightarrow h$	$G2E : g \rightarrow e$	$H2I : h \rightarrow i$

The planning graph is $\{a\} \rightharpoonup \{A2BC, A2D\} \rightharpoonup \{a, b, c, d\} \rightharpoonup \{C2E, D2E, D2F\} \rightharpoonup \{a, b, c, d, e, f\}$.

The solutions are $A2BC; C2E$ and $A2D; D2E$. If we remove service $C2E$, we still have another solution. If we remove both $C2E$ and $D2E$, the graph is broken, because e is not produced by any services. We have the following partial planning graph, where e is unsatisfied: $\{a\} \rightharpoonup \{A2BC, A2D\} \rightharpoonup \{a, b, c, d\} \rightharpoonup \{D2F\} \rightharpoonup \{a, b, c, d, \underline{e}, f\}$.

If we do a replanning, a new planning graph is built from scratch: $\{a\} \rightharpoonup \{A2BC, A2D\} \rightharpoonup \{a, b, c, d\} \rightharpoonup \{D2F\} \rightharpoonup \{a, b, c, d, f\} \rightharpoonup \{F2G, F2H\} \rightharpoonup \{a, b, c, d, f, g, h\} \rightharpoonup \{G2E, H2I\} \rightharpoonup \{a, d, c, d, f, g, h, e, i\}$. This graph contains a solution $A2D; D2E; F2G; G2E$.

The repair algorithm developed in this paper does not build the full planning graph as above, but tries to fix the partial planning graph while searching for a solution. In order to produce the unsatisfied e, we first add a service $G2E$ into the partial graph, which results: $\{a\} \rightharpoonup \{A2BC, A2D\} \rightharpoonup \{a, b, c, d, g\} \rightharpoonup \{G2E, D2F\} \rightharpoonup \{a, b, c, d, e, f\}$.

Next, we add a service $F2G$ to satisfy g: $\{a, f\} \rightharpoonup \{A2BC, A2D, F2G\} \rightharpoonup \{a, b, c, d, g\} \rightharpoonup \{G2E, D2F\} \rightharpoonup \{a, b, c, d, e, f\}$.

We continue this process until the following graph is built: $\{a\} \rightharpoonup \{A2D\} \rightharpoonup \{a, d\} \rightharpoonup \{D2F\} \rightharpoonup \{a, f\} \rightharpoonup \{A2BC, A2D, F2G\} \rightharpoonup \{a, b, c, d, g\} \rightharpoonup \{G2E, D2F\} \rightharpoonup \{a, b, c, d, e, f\}$.

Now, we have no unsatisfied preconditions and goals. Therefore, we can get a solution $A2D; D2F; F2G; G2E$. We can see that the repaired graph is simpler than the full planning graph. Our goal is to find at least one solution.

4   Evaluation Criteria

We want to evaluate the efficiency and quality of replanning and repair in generating new Web service composition solutions. We use three kinds of criteria: execution time, the quality of the new composition, and plan stability.

The composition time

It is the time to get the first feasible solution or report nonexistence of it. It is possible that after removing some services or adding new goals, no solution exists to the new problem. For replanning, since we solve the new problem using planning graphs, we are able to know the existence of solution. For repair, as we solve it using heuristics (see the next section), the algorithm may report no solution in a case where solutions exist.

The quality of the new solution

A solution is better if it invokes less services and if it involves less time steps. For planning graphs, the time steps are the levels of action layers. In one action layer, the services can be executed in parallel. If less services are invoked, the cost of the solution is lower, assuming each service has the same execution cost.

Plan Stability

Plan stability is defined in Defition 9 as in [7]:

Definition 9. *Given an original plan, π_0, and a new plan π_1, the difference between π_0 and π_1, $D(\pi_0, \pi_1)$, is the number of actions that appear in π_1 and not in π_0 plus the number of actions that appear in π_0 and not in π_1.*

We are interested in achieving plan stability because we prefer that the new plan is similar to the original one. In the Web service composition context, this means that we can keep our promise to the business partners in the original plan.

5   Repair the Broken Planning Graph to Generate a Solution

Due to the changing world, the available service set and the goal set are updated. In this paper, we are interested in finding a feasible solution fast, rather than an optimal solution. If more services are available or less goals needed to satisfy, the original planning graph can still give a solution. Therefore, we focus on the cases when services are removed or more goals are added, in which cases, the original planning graph becomes only partially valid.

Our idea is to "grow" the partial planning graph again in order to obtain a feasible solution. Our approach is not to build an entire planning graph again, but to fast find a solution. During the "growing" process, we heuristically search into a direction that can satisfy the broken goals and preconditions, as well as making use of the existing partial graph. Our strategy is as follows. Assume

- w: a service whose inputs are $in(w)$ and outputs are $out(w)$;
- G: a partial planning graph;
- $\tilde{g}$: a set of unimplemented goals;
- BP: a set of unsatisfied preconditions (inputs) of some services in G;
- D_U^{in} is the initial proposal level noted as P_0; the first action level is noted as A_1; the last proposition level P_n which is also the D_U^{out} level.

Assume we want to add an action w to the highest action level n, the evaluation function is:

$$f(G,w) = |\tilde{g} \cap out(w)| * 10 + |P_{n-1} \cap in(w)| - |in(w) - P_{n-1}| - e(w,G) \qquad (5)$$

Where $|\tilde{g} \cap out(w)|$ is the number of unimplemented goals that can be implemented by w. The coefficient 10 is the weight of this term. It shows that to satisfy the goals is more important than the other needs represented by the following terms; $|P_{n-1} \cap in(w)|$ is the number of the inputs of w that can be provided by the known parameters at the level P_{n-1}; $|in(w) - P_{n-1}|$ is the number of the inputs of w that **cannot** be provided by the known parameters at the level P_{n-1}. This set needs to be added into BP, if w is added. $e(w,G)$ is the number of the actions in G that are exclusive with w.

Assume we want to add an action w to action level m, and m is not the goal level, the evaluation function is:

$$f(G,w) = |\tilde{g} \cap out(w)| * 10 + |P_m \cap BP \cap out(w)| + |P_{m-1} \cap in(w)| - |in(w) - P_{m-1}| - e(w,G) \qquad (6)$$

Compared to equation 5, the above equation added term $|P_m \cap BP \cap out(w)|$ which is the number of the broken propositions in level P_m that can be satisfied by the outputs of w.

Algorithm 1 uses heuristics to search for a repaired composition. The algorithm starts from the goal level. It tries to satisfy the unimplemented goals first (Line 2 to 8). When new services are added into the partial planning graph, its precondition may not be satisfied. The unsatisfied preconditions are added to BP to be satisfied at a lower level (Line 8, 17 and 29). This process goes from the goal level toward the initial proposition level (Line 10 to 18). It is possible that after adding actions at A_1, we still have some broken preconditions. In this case, we need to add new levels (Line 19 to 32). We use BPH to record the history of preconditions we have satisfied. If $BPH \cap BP \neq \emptyset$, that means some precondition broken for the second time. It is a deadlock situation. We stop the search.

Algorithm 1 is a greedy search algorithm. It does not generate a full planning graph, but rather, to fast fix the broken graph and obtain a feasible solution. It is possible that Algorithm 1 does not find a solution while there is one. However, repair

---

Algorithm 1. Search for a repair plan: Repair(W, G, BP, $\tilde{g}$)

- Input: W, G, BP, $\tilde{g}$ defined as before
- Output: either a plan or `fail`

1: result = `fail`
2: **while** $\tilde{g} \neq \emptyset$ **do**
3:     select an action w with the best $f(G,w)$ according to equation 5
4:     **if** w does not exist **then** break
5:     add w to G at action level A_n
6:     add $out(w)$ at proposition level P_n
7:     remove $\tilde{g} \cap out(w)$ from $\tilde{g}$
8:     add $in(w) - P_{n-1}$ to BP
9: **if** $\tilde{g} \neq \emptyset$ **then** return result
10: **for** $m = n - 1; m > 0; m - -$ **do**
11:     **while** $BP \cap P_m \neq \emptyset$ **do**
12:         select an action w with the best $f(G,w)$ according to equation 6
13:         **if** w does not exist **then** break
14:         add w to G at action level A_m
15:         add $out(w)$ at proposition level P_m
16:         remove $P_m \cap BP \cap out(w)$ from BP
17:         add $in(w) - P_{m-1}$ to BP
18:     **if** $BP \cap P_m \neq \emptyset$ **then** return result
19: $BPH = BP$
20: **while** $BP \neq \emptyset$, $W \neq \emptyset$ **do**
21:     insert an empty proposition level P_1 and empty action level A_1
22:     $P_1 = P_0 - BP$
23:     **while** $BP \cap P_1 \neq \emptyset$ **do**
24:         select an action w with the best $f(G,w)$ according to equation 6
25:         **if** w does not exist **then** break
26:         add w to G at action level A_1
27:         add $out(w)$ at proposition level P_1
28:         remove $P_1 \cap BP \cap out(w)$ from BP
29:         add $in(w) - P_0$ to BP
30:         remove w from W
31:     **if** $BP \cap P_1 \neq \emptyset$ **then** break
32:     **if** $BP \cap BPH \neq \emptyset$ **then** break **else** add BP to BPH
33: **if** $BP \neq \emptyset$ **then** return result
34: **else** {repaired successfully}
35:     result = backwardsearch(G)
36: return result

---

can be faster than replanning, which keeps the quality of solutions (cf. Section 6). It is also possible that Algorithm 1 generates a graph that contains multiple solutions. Therefore, a `backwardsearch()` function returns the first obtained solution. We know that a solution can be found from the graph made of the originally broken plan and the newly added services. `backwardsearch()` is on this very small graph, thus fast.

6   Empirical Evaluation

6.1   Data Set Generation

We use the Web Service Challenge 2009 [2] platform and its tools in our experiments. The platform can invoke the composition algorithm as a Web service and evaluate composition time. The data generator generates ontology concepts in OWL and a set of Web services interfaces in WSDL which use the concepts. Given the number of services and time steps in a solution, the generator algorithm generates a number of solutions around these numbers first, and then randomly generates a set of Web services. A validation tool can verify the correctness of the result BPEL file by simulating the process to see whether the query parameters are produced. The 2009 challenge supports QoS features which are not used our experiments. We use two data sets. The first one is relatively small with 351 available Services. These Services use 2891 Parameters. These Parameters are selected from 6209 Things. And these Things are instances of 3081 Concepts. The second one is larger with 4131 available Services, which involve 45057 Parameters, 6275 Things and 2093 Concepts. Concepts correspond to the classes defined in OWL. Things are instances of Concepts and are used as the Parameter names (params) for Services. Services are described in a WSDL file. Each Service has exactly one port type and each port type has one input message and one output message. Notice that subsumption is modeled in OWL and is used during composition. Negative effects are not modeled in this kind of data sets.

6.2   Implementation and Experiments Set Up

The purpose of this experiment is to compare our repair algorithm with the replanning algorithm which is a standard planning graph algorithm in case of updated problem. The replanning serves as a baseline algorithm. We want to check whether repair algorithm can be better than replanning and, if so, under which conditions. We use four criteria to evaluate the performance of algorithms presented in Section 4. We conduct two experiments. In both of the experiments, a solution is chosen to be the original solution. In Experiment 1, a certain percentage of available services, from 3% to 24%, is randomly removed. If the original solution is broken, we start to repair it. In Experiment 2, a number of services are randomly removed from the chosen solution. And a repair is thus needed. The situation of adding new goals is similar to the situation where some goals are not satisfied due to service removal. Thus we only consider service removal in our experiments.

Our implementation includes some technical details that cannot fully be discussed here. First, we apply several indexing techniques for expediting composition process. For each service, we use a hash table to index all the possible concepts (defined in a OWL document) that the service takes as inputs or outputs. The subsumption hierarchy is thus "flattened" in this step so we do not need to consider semantic subsumption during the planning process. Second, we also use a hash table to store the mapping relationships between each semantic concept and all services that can accept that concept as one of their inputs. This is similar to the "reverse indexing approach" introduced in [28]. It allows us to search the invokable Web services from the known concepts very fast - a simple join operation among related rows of the index table instead of checking the precondition of each service in the repository.

6.3 Results

The following comparison of performance is recorded in the cases that both repair and replanning algorithms can find solutions. Each data point is obtained from the average of five independent runs.

Fig 2 to Fig 5 show the results from Experiment 1. From Fig 2, we can see that the replanning composition time is slightly decreasing when more Web services are removed. It is because the problem is smaller when less Web services are available. However, it is more difficult to repair the plan. Therefore, the repair composition time increases in such a case. However, after a certain percentage (around 20%), the repair composition time descreases. This is because the problem becomes simpler to solve and also because we are less committed to the original composition. Please notice that repair may not find existing solutions.

Fig 3 and Fig 4 show the number of services and the number of levels in a composed plan. The plot for repair is rather flat. Our explanation is that our repair algorithm does not work well when the number of levels are over 10. This is because it is a greedy search algorithm and the successful rate is lower in more level cases. As

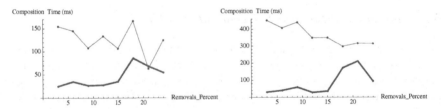

Fig. 2 Composition time with data set 1 (left) and data set 2 (right) in Experiment 1 (repair - thick line, replanning - thin line)

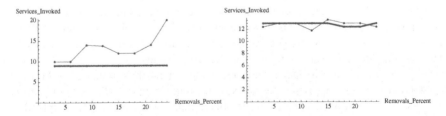

Fig. 3 Number of services in the solution with data set 1 (left) and data set 2 (right) in Experiment 1 (repair - thick line, replanning - thin line)

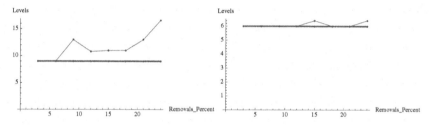

Fig. 4 Number of levels in the solution with data set 1 (left) and data set 2 (right) in Experiment 1 (repair - thick line, replanning - thin line)

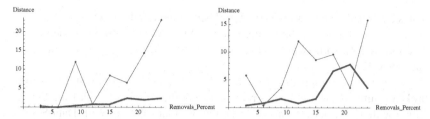

Fig. 5 Distance to the original plan with data set 1 (left) and data set 2 (right) in Experiment 1 (repair - thick line, replanning - thin line)

we do not count unsuccessful cases, we end up showing the cases where the levels are below 10 and very flat. Fig 3 and Fig 4 show that when repair finds a solution, the quality of the solution is pretty good.

Finally, Fig 5 shows that the solution computed with the repair algorithm can be more similar to the original plan than the solution computed with the replanning algorithm.

A brief discussion on the failure rate of Experiment 1. Algorithm 1 can fail to find a solution when one exists due to two reasons. First, Algorithm 1 is a greedy search without backtracking. It is possible that some data traps the algorithm into a dead end. However, due to the nature of the Web Service Challenge data, this is unlikely to happen. It is simply because the data set does not have dead ends. Second, after the removal, the original solution is not repairable. This happens when a non-replaceable service in the original solution is removed. Algorithm 1 has no ability to detect this or switch to some other possible solutions. For the solution chosen to be the original solution, 3 of 10 services in the solution are non-replaceable, i.e. if either of them is removed, Algorithm 1 cannot repair the solution successfully. Therefore, the failure rate equals to the probability of either of the three is removed when we remove $x\%$ services from 351 avaliable services. Mathematically, it is a problem to calculate the probability of the occurrence of a black ball out of a box of r red balls and b black balls, if we draw $n \leq r$ times. It is easier to calculate the probability of getting all red balls out of the n draws, which can be calculated as follows: $P(r|n) = P(r_1)P(r_2|r_1)P(r_3|r_2,r_1)...P(r_n|r_1,...,r_{n-1}) = r/(r+b) \cdot (r-1)/(r+b-1) \cdots (r-n+1)/(r+b-n+1) < (r/(r+b))^n$, where r_i

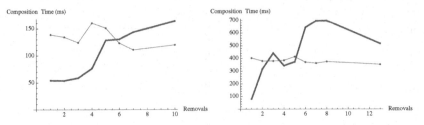

Fig. 6 Composition time with data set 1 (left) and data set 2 (right) in Experiment 2 (repair - thick line, replanning - thin line)

denotes a red ball comes at the *i*-th draw. Therefore the failure rate is $1 - P(r|n) > 1 - (r/(r+b))^n$. If $n > r$, the failure rate is 100%. As one example, when the removal rate is 20% with Data Set 1, *i.e.*, $n \approx 70$, $r = 348$, $b = 3$, the failure rate is greater than 45%.

Fig 6 shows the composition time from Experiment 2. Experiment 2 tries to fix a broken solution (services beings removed from the solution). With Data Set 1, a solution with 10 services is chosen to be the original solution. With Data Set 2, a solution with 14 services is chosen. Since the number of removed services are not big among the whole available services, the composition time for the replanning algorithm does not change a lot for all the runs, because the composition problems on each run are of the same difficulty. When a small number of services are removed, repair is faster than replanning. But this becomes more difficult as more services are removed. Therefore, at one moment, repair takes longer than replanning. However, if services from an original solution are almost completely removed, then repair is like searching for a new plan, which can become faster. This shows in the last data point at the right figure in Fig 6. The comparison of solution quality and distance to original solution is similar as in Experiment 1 and is omitted.

A brief discussion on the failure rate of Experiment 2. Please reference to the discussion on the failure rate for Experiment 1. The principle is the same, except that in Experiment 2, the removal is within the original solution. Therefore, the numbers of the red balls and the black balls are close, for example, for Data Set 1, $r = 7$, $b = 3$. If one service is removed, the failure rate is $3/10 = 30\%$. If $n > 7$, the failure rate is 100%. Fig 6 shows the composition time when both replanning and repair algorithms can successfully find a solution. One can see that with Data Set 1, no data point exists at $n = 8$ and $n = 9$ in the left figure in Fig 6, it is because the original plan is not repairable. When $n = 10$, the whole original plan is removed. Algorithm 1 finds another solution. The analysis for Data Set 2 is similar.

7 Related Work

Web service composition has been studied under various assumptions and models. In case where it is impossible to get information on the internal behavior of services, we can consider only how to connect the input/output parameters of services. We can further assume *stateless services* and use *synchronous communication* (request-response style). This is what we call ***data driven problems*** with automatic services.

This WSC problem can be modeled as a planning problem [17] by mapping a service (operation) to an action, the service input parameters as preconditions of the action, and the service output parameters as positive effects of the action. If a parameter can be used only once, it becomes a negative effect. For some composition requirement (D_U^{in}, D_U^{out}) and a set of actions for services $W = \{w_1, w_2, \ldots, w_n\}$, if one can find a planning structure with D_U^{in} as initial state and D_U^{out} as goal, then the WSC problem can be resolved by the network of W taking D_U^{in} as input and producing D_U^{out} as results. Many planning algorithms can be used to solve the above planning problem, *e.g.*, heuristic search in AI [17] or linear programming [29].

WSC problems also consider semantics defined in languages such as OWL-S and WSMO. A review of some early approaches on planning with description logics can be found in [11]. Due to length limitation, this aspect is not reviewed here.

If we assume services have an internal behavior, *i.e.*, a service is a stateful process, then we need to consider asynchronous communication since we may not get a response immediately after a request message. The response can be produced by some internal nondeterministic process. Usually, the complete internal behavior of services is not exposed. However, behaviors can be partially observed through input and output messages, *e.g.*, using abstract BPEL. We call this ***process driven problems***.

Pistore and his colleagues' work [20] can handle asynchronous communication and the partial observable internal behavior of the candidate services. Their solution is based on belief systems and is close to controller synthesis. The condition to build belief systems is too strong and a large portion of the systems is not controllable (*i.e.*, not deadlock free). Van der Aalst and his colleagues try to solve the same problem [22] using Operation Guide Lines (OGL). While it is often impossible to know the internal behavior of a service, this service may expose its behavior in some way anyway, *e.g.*, in abstract BPEL. An OGL for a service abstracts the service controllable behavior such that, if a partner service can synchronize with the service OGL without deadlocks, then the partner service can also synchronize with the service itself. The difficulty is how to abstract OGL.

The major difference between our work and the above mentioned ones is that we tackle the problem of adapting service compositions to the changing world. This issue is also addressed by replanning in domain independent AI planning. Replanning normally has two phases: a first one determines the parts where the current plan fails and the second one uses slightly modified standard planning techniques to replan these parts [14]. Slightly modified planning techniques normally mean the use of heuristics of recorded decision and paths in searching for actions. In theory, modifying an existing plan is no more efficient than a compete replanning in the worst case [23]. However, we have shown that in some cases, *e.g.*, when the number of services to replace is small wrt. the total number of services used in a composition, repair is a valuable alternative to replanning in terms of computation time. One could notice that indeed both replanning and repair took little time. Still, we advocate that repair also has better results in terms of distance to the original plan, which is an important parameter for the end-user that is to use the repaired service composition. Complementarily to generated data using using the Web Service Challenge platform [2], it would be interesting to run experiments on data retrieved from existing service repositories. An alternative is to use statistical approaches in the generation of the simulation data as done in [13].

The DIAMOND project [1] also studies the repair problem. [8] presents a method of composing a repair plan for a process. It predefines substitution and/or compensation actions for an action in a process. [8] is based on a logical formalization which can be processed by modern logical reasoning systems supporting disjunctive logic programming. Similar work can be found in the replanning area as well, for example O-Plan [5]. Whenever an erroneous condition is encountered, the execution of the plan is stopped and some predefined repair actions are inserted and executed. In contrast, our paper is not about fault diagnosis or exception handling. We change the original plan with available Web services to make it valid again.

8 Conclusions and Future Work

Service compositions have to be adapted whenever the composition context, *i.e.,* available services and composition requirements, change. In this paper we have set up service composition in the AI planning domain and we have proposed a planning graph repair algorithm focused at obtaining an updated service composition solution fast rather than obtaining all possible modified service composition solutions. We have proposed composition adaptation comparison criteria, namely composition time, plan quality, and distance between the original and the modified plan. We have then used these criteria to compare our repair algorithm with replanning which is the standard planning technique for plan modification. We observed that when services are removed on a percentage basis of the whole service set, and when available services are many and similar as in Experiment 1, our repair algorithm can be faster than the replanning one. The repaired plan can be as good as the one obtained with replanning but is more similar to the original plan. In the case where one wants to fix an existing plan in which some services are removed, as in Experiment 2, the repair algorithm is faster than replanning only when the number of removed services is small. At a certain point, repair then gets slower than replanning. This suggests there is no silver bullet in adapting plans and that different algorithms should be selected depending on the kind and degree of change.

There are different perspectives to our work. First, we can improve the repair algorithm following existing work in replanning [9] [23] that propose, upon disappearance of some service(s), to remove even more services in order to jump out of local extrema in the search for a new plan. A second perspective is relative to the assumptions we made on services and composition requirements. Taking into account behavioral properties, *e.g.,* stateful services and conversation-based requirements, would increase the expressiveness of our approach as far as Web services are concerned.

Acknowledgements. This work is supported by project "Building Self-Manageable Web Service Process" of Canada NSERC Discovery Grant, RGPIN/298362-2007 and by project "PERvasive Service cOmposition" (PERSO) of the French National Agency for Research, ANR-07-JCJC-0155-01.

References

1. Diamond project homepage, http://wsdiamond.di.unito.it/
2. Web service challenge 2009 web site (2009), http://ws-challenge. georgetown.edu/wsc09/index.html
3. Beauche, S., Poizat, P.: Automated Service Composition with Adaptive Planning. In: Bouguettaya, A., Krueger, I., Margaria, T. (eds.) ICSOC 2008. LNCS, vol. 5364, pp. 530–537. Springer, Heidelberg (2008)
4. Bleul, S.: Web service challenge rules (2009), http://ws-challenge. georgetown.edu/wsc09/downloads/WSC2009Rules-1.1.pdf
5. Drabble, B., Dalton, J., Tate, A.: Repairing Plans On-the-fly. In: NASA Workshop on Planning and Scheduling for Space (1997)
6. Dumas, M., Benatallah, B., Motahari-Nezhad, H.R.: Web Service Protocols: Compatibility and Adaptation. Data Engineering Bulletin 31(3), 40–44 (2008)
7. Fox, M., Gerevini, A., Long, D., Serina, I.: Plan Stability: Replanning versus Plan Repair. In: Long, D., Smith, S.F., Borrajo, D., McCluskey, L. (eds.) Proc. of ICAPS, pp. 212–221. AAAI, Menlo Park (2006)
8. Friedrich, G., Ivanchenko, V.: Model-based repair of web service processes. Technical Report 2008/001, ISBI research group, Alpen-Adria-Universität Klagenfurt (2008), https://campus.uni-klu.ac.at/fodok/veroeffentlichung. do?pubid=67566
9. Gerevini, A., Serina, I.: Fast planning through greedy action graphs. In: AAAI/IAAI, pp. 503–510 (1999)
10. Ghallab, M., Nau, D., Traverso, P.: Automated Planning: Theory and Practice. Morgan Kaufmann Publishers, San Francisco (2004)
11. Gil, Y.: Description Logics and Planning. AI Magazine 26(2) (2005)
12. Hashemian, S.V., Mavaddat, F.: Automatic Composition of Stateless Components: A Logical Reasoning Approach. In: Arbab, F., Sirjani, M. (eds.) FSEN 2007. LNCS, vol. 4767, pp. 175–190. Springer, Heidelberg (2007)
13. Kambhampati, S., Parker, E., Lambrecht, E.: Understanding and Extending Graphplan. In: Steel, S., Alami, R. (eds.) ECP 1997. LNCS, vol. 1348, pp. 260–272. Springer, Heidelberg (1997)
14. Koenig, S., Furcy, D., Bauer, C.: Heuristic search-based replanning. In: Ghallab, M., Hertzberg, J., Traverso, P. (eds.) Proc. of AIPS, pp. 294–301. AAAI, Menlo Park (2002)
15. Marconi, A., Pistore, M.: Synthesis and Composition of Web Services. In: Bernardo, M., Padovani, L., Zavattaro, G. (eds.) SFM 2009. LNCS, vol. 5569, pp. 89–157. Springer, Heidelberg (2009)
16. Melliti, T., Poizat, P., Ben Mokhtar, S.: Distributed Behavioural Adaptation for the Automatic Composition of Semantic Services. In: Fiadeiro, J.L., Inverardi, P. (eds.) FASE 2008. LNCS, vol. 4961, pp. 146–162. Springer, Heidelberg (2008)
17. Oh, S.-C., Lee, D., Kumara, S.: Web Service Planner (WSPR): An Effective and Scalable Web Service Composition Algorithm. Int. J. Web Service Res. 4(1), 1–22 (2007)
18. Oh, S.-C., Lee, D., Kumara, S.: Flexible Web Services Discovery and Composition using SATPlan and A* Algorithms. In: Proc. of MDAI (July 2005)

19. Peer, J.: Web Service Composition as AI Planning – a Survey. Technical report, University of St.Gallen (2005)
20. Pistore, M., Traverso, P., Bertoli, P.: Automated Composition of Web Services by Planning in Asynchronous Domains. In: Proc. of ICAPS, pp. 2–11 (2005)
21. Seguel, R., Eshuis, R., Grefen, P.: An Overview on Protocol Adaptors for Service Component Integration. Technical report, Eindhoven University of Technology. BETA Working Paper Series WP 265 (2008)
22. van der Aalst, W.M.P., Mooij, A.J., Stahl, C., Wolf, K.: Service Interaction: Patterns, Formalization, and Analysis. In: Bernardo, M., Padovani, L., Zavattaro, G. (eds.) SFM 2009. LNCS, vol. 5569, pp. 42–88. Springer, Heidelberg (2009)
23. van der Krogt, R., de Weerdt, M.: Plan Repair as an Extension of Planning. In: Biundo, S., Myers, K.L., Rajan, K. (eds.) Proc. of ICAPS, pp. 161–170. AAAI, Menlo Park (2005)
24. W3C. Owl web ontology language overview (2004),
    http://www.w3.org/TR/owl-features/
25. W3C. Semantic annotations for wsdl and xml schema, sawsdl (2007),
    http://www.w3.org/TR/sawsdl/
26. W3C. Web services description language (wsdl) version 2.0 (2007),
    http://www.w3.org/TR/wsdl20/
27. Yan, Y., Dague, P., Pencolé, Y., Cordier, M.-O.: A Model-Based Approach for Diagnosing Fault in Web Service Processes. Int. J. Web Service Res. 6(1), 87–110 (2009)
28. Yan, Y., Xu, B., Gu, Z.: Automatic Service Composition Using AND/OR Graph. In: Proc. of CEC/EEE, pp. 335–338. IEEE, Los Alamitos (2008)
29. Yoo, J.-W., Kumara, S., Lee, D., Oh, S.-C.: A Web Service Composition Framework Using Integer Programming with Non-functional Objectives and Constraints. In: Proc. of CEC/EEE, pp. 347–350 (2008)

Remote Automated User Testing: First Steps toward a General-Purpose Tool

Chandan Sarkar, Candace Soderston, Dmitri Klementiev, and Eddy Bell

Abstract. In this paper we explore options for conducting remote, unattended usability tests to enable users to participate in their own environments and time zones, including multiple users' participating at the same time in remote, unattended studies. We developed a general purpose tool, and code-named it "TCA" (for "Total Cost of Administration") to catalog and analyze database administrators' behavior within software. In this paper, we present example findings from the data collected over a period of 6 months. We analyzed users' deviations from the best paths through the software, in addition to collecting traditional measures such as time on task, error rate, and users' perceptions, including satisfaction level. Further, we explore how this type of tool offers particular promise in benchmark studies, also collecting the ideal best-path performance that assumes error-free, expert user behavior, to compare to more real-world data collected initially in laboratory tests or over time in longitudinal field studies.

Keywords: Remote testing, unattended testing, automated testing, benchmarking.

1 Introduction

Shaping technologies to fit the needs and expectations of human beings for future use is a challenge for all who work in technology creation and development. Part of what makes this a challenging problem is the art associated with understanding and predicting the users' task goals, skills, and what delights them over time. There are many aspects to this art, and understanding when and where users

Chandan Sarkar
Michigan State University
College of Communication Arts & Sciences, East Lansing, Michigan, USA
e-mail: sarkarch@msu.edu

Candace Soderston
Microsoft Corporation
One Microsoft Way, Redmond, Washington, USA
e-mail: csoders@microsoft.com

Dmitri Klementiev and Eddy Bell
i-PEI LLC
25519 176th Street SE, Monroe, Washington, USA
e-mail: dklem@microsoft.com, eddy@i-pei.com

Roger Lee (Ed.): SERA 2010, SCI 296, pp. 37–50, 2010.
springerlink.com   © Springer-Verlag Berlin Heidelberg 2010

experience difficulties with a new or existing user interface is essential to improving the design of products. Good usability tests today analyze how users interact with a design, the path they take through software in order to achieve a relevant task, the time it takes them to complete the task, the errors they make, and their perception, including satisfaction level.

Working with users to simulate their work in a usability laboratory environment, to evaluate design alternatives and bring the resulting understandings back to a larger group of people associated with building a product, is a well accepted process, used now by user researchers and designers for decades [6,19,23,24,25]. A typical lab-based usability study involves close observation of a selected number of users with elaborate set-ups for video-taping and manual data logging, time synched with the video capture. Typically, a small number of representative users (e.g., six to eight) [14,15,33] are recruited for a study. The sessions with the study participants are conducted one at a time (linearly, not in parallel), where they perform a series of tasks using an application or prototype. One or more observers in an adjacent "control" room, out of the participants' way, actively view and monitor their performance through a one-way glass window and projection of the user's screen. They record the time to complete each task successfully, and any behaviors and errors that block task progress [8]. This information is then used to develop a list of "usability issues" or potential problem areas to improve in the application [29].

This traditional lab-based usability testing is very useful and provides key insight, however, it can be time-consuming and expensive when a desired target audience is geographically dispersed. The cost of recruiting participants across geographies and bringing them physically into a usability lab can be high and logistically challenging.

Remote testing offers some interesting alternatives to a physical "lab-based" approach, providing test moderators and participants more flexibility and, potentially, a more natural experience (in the users' environment). In addition, conducting remote studies opens up the possibility of running tests "unattended", where users log in at their convenience, across time zones, and complete the tasks through structured and sequenced instructions presented electronically. We were very interested in this aspect, to both widen the participant base and to reduce the overhead burden/time in conducting tests. We also recognized that for unattended remote study the data analysis would be more challenging and could be more time consuming, as the test moderator would not be there to interpret and manually log impressions about the users' activities, errors, and so on.

In this study, we set out to explore options for conducting remote, *unattended* usability tests that would enable users to participate in their own environments and on their own time, and would enable multiple users to do the tests at the same time. We describe what we found regarding the feasibility and the perspectives of user experience researchers and information architects. We cover key aspects of the tool we created to help us, and present what we found in pilot testing this tool. We explore different analysis patterns and wrap up with a discussion of results and future implications.

2   Related Work

In the past, the classic GOMS model (Goals-Operators-Methods-Selections) was used to develop computer based tools to support interface design for the analysis of tasks [4]. For example, researchers experimented with tools like GLEAN and CHIME which attempt to recognize higher order task chunks from keystroke and mouse action data, and to generate quantitative predictions from a supplied GOMS model. [1,10].

The GOMS family tools can be useful, but they don't predict the time taken by a user to complete tasks (e.g., including visual search time and "think" time), and where in the UI users will likely make errors.[10]

In order to address the above challenges, researchers have explored automated data logging to collect "click stream" data of the user's action sequences [3,7,35,36]. For example, automated data logging was used during the design of the User Interface Management Systems (UIMS) [17]. UIMS was one of the first published instances where interface modifications were made to track and collect user behavior (the code was instrumented) and where integrated support was provided for an evaluator's data collection and analysis activities. Usability engineers and researchers have refined and extended this early approach to remote inspection, remote questionnaire survey, remote control evaluation, and video conferencing to support similar needs and requirements [8,9].

To enable attended remote testing which does not lose the above aspects, various usability testing tools for remote data logging and analysis have been produced and are available on the market today (examples: Morae, Ovo, Silverback, Uservue,)[12,18,21 32]. Though not free of shortcomings, these tools provide support to usability engineers for recording, observing, managing, and analyzing data.

Automated data collection via plug-ins to browsers has gained popularity in terms of remote usability testing for web-sites. Gathering click stream information for testing enables more participants outside a usability lab to be involved, in their own environments, doing their own tasks, but this approach lacks the qualitative information that can yield additional insights, such as spoken comments, physical reactions, facial expressions, gestures and other non-verbal cues observable in the lab [34].

In recent years several efforts have been made to design efficient and useful benchmark kits. Benchmarking tools are useful to compare different hardware configuration, different vendor software and different releases in real world. Earlier studies have demonstrated the design of tools such as the Automated Summative Evaluation (ASE) tool [36] and the Direction Mapping and Sequence Chart (DMASC) [27] tool for benchmarking and information visualization purposes.

Various open source free benchmarking tools exist, including PolePosition, a benchmark test suite which compares the database engines and object-relational mapping technology for same and different databases[20], BenchmarkSQL, a Java Database Connectivity (JDBC) benchmark tool for testing online transaction processing (OLTP) operations [2], and Swingbench, a stress test tool for Oracle databases (9i,10g,11g) [28]. In addition to the free tools, commercial tools are also available, for example the Standard Performance Evaluation Corporation's (SPEC) jAppServer Benchmark Kit and Official Oracle Benchmark Kit for database bench marking [22][16]. Most of these tools are good for defined bench marking operations but not geared for an end-to-end spectrum of tasks, including the user experience aspects in addition to the system processing times.

The fact that these user test and system benchmark test tools exist reflects the importance and good attention given to automating or making these test efforts easier than they would otherwise be. While there are good things about these tools, more work also needs to be done, for example to minimize latency delays setup processes, and disk space for installation [31]. There are far more studies of why these attempts have failed rather than how to succeed in doing remote user testing. This was our starting point, and our objective related to how we could take this further, to make remote, automated testing a normal part of our usability engineering practices.

In order to pilot test our remote automated testing capability, we looked for tools and found a good one to start with and then continue to develop it as needed. Collaborating with an architect in our larger team, we created a tool called the "Total Cost of Administration" (TCA) tool. The TCA name came from its initial purpose to benchmark system workload or productivity, assuming the ideal, error-free database administrator or user behaviour, to capture system performance throughput as a "total cost" time measure. We saw promise in extending this tool for use in real user evaluations, with "unguided" behaviour that can differ widely from ideal, error-free behaviour.

TCA was built on the top of the standard Microsoft Active Accessibility (MSAA) framework [13] that enables standard software to be used by special needs audiences. Because of the standard protocol of MSAA (MAT today), we were fortunate that we could build upon a tool that already existed and was used by test organizations. This tool, called the Record and Playback Framework [RPF], among other things assesses code readiness to the MSAA standard, to ensure that standard assistive technologies can be written to interpret user behavior and selections, for example, to be able to read out the appropriate label and prompts to blind users when they are on a particular field. To do this, RPF tracks user's action sequences and, to be able to record and play it back, collects the time-stamp for each action, enabling us to assess the duration of each task, the time on task, and the path followed while performing the task.

TCA was a low cost investment from our side in terms of code and cost since we could add user task instructions and then post-task filters and interpreters into TCA, to run on top of the already-existing RPF framework. Because in a remote, unattended user study, we have no access to information on what users are doing behind the scenes, TCA enables us to collect these data to analyze later. The TCA or RPF code could also be used in field studies, in the user's "real world" operations, in place of asking people to keep a diary of their software usage, obstacles, and delight factors.

We think of TCA now as an automated UI data logger and benchmarking tool, built from the ground up in C# and XML to collect user action sequences. TCA is designed to look for particular action sequences and clicks of the user, to identify

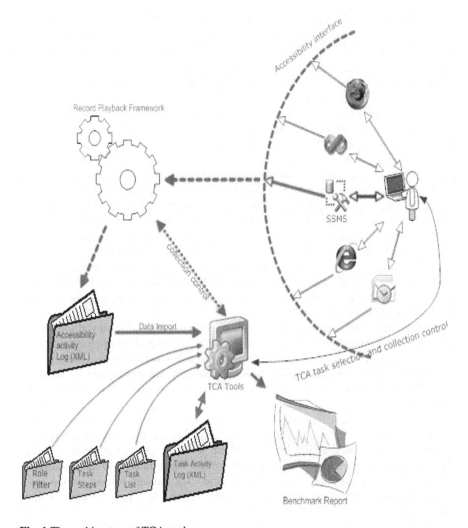

Fig. 1 The architecture of TCA tool

certain data-collection and benchmark patterns of interest. Test administrators can track all or look for only some particular patterns, depending on the objective of the study.

RPF is made up of a Record Module and a Playback Module. The Record module is made up of a collector, an aggregator and a code generator module. The collector module is an aggregation of filters, which determine how much data is collected. The collector collects the data from the MSAA interfaces, capturing the UI selections and mouse actions. The aggregator eliminates redundant data by aggregating it into more meaningful chunks: for example, mouse-button-1-down, move, and mouse-button-1-up actions in sequence could be aggregated into a single mouse drag/drop action "chunk". The code generator acts as an interface between the Record and Playback modules. In the playback module it interprets the recording and replicates the UI activity. The TCA, the RPF Framework and the accessibility logs form an inner loop, processing the stream of inputs and presenting that in a more human-readable format (See figure 1).

The current version of the TCA tool is designed in two separate formats - one for usability studies as an automated data logger and analysis tool, and other one for a guided step-by-step ideal path pure benchmark tool for best performance on tasks, including error-free user behavior and system performance/throughput time. Both versions generate XML and plain text output. The usability study version is instrumented to collect user actions and the task step-by-step guidance is suppressed (silent). The other version provides step-by-step instruction for a benchmark test that assumes error-free user performance to collect the best path time on task. The setup and configuration of the usability study and guided best path versions of the tool are relatively simple and require only few minutes to complete.

We coupled the data logger tool along with the "task book" to initiate the recording of the users' action sequence. The task book is designed as a small browser window across the top of the main screen. The task book presents the task goals to the user, and captures user input, feedback, and Likert scale ratings for each task. The main window, which fills the majority of the screen, is used for the application being tested.

Any recorded task which is benchmarked can be saved and retrieved for future reference. The compare functionality provides a measure of comparison and deviations between participants and multiple approaches for doing the same tasks. During performing any benchmark task the collector can be stopped by the end user. The tool provides a method to document participants' comments for each task steps.

In order to test the effectiveness of the TCA tool for remote usability testing, we pilot tested the tool in a "virtual remote lab" configuration and ran a series of traditional monitored usability lab tests where we observed and manually logged the data as well as collected the TCA data. To test the remote unattended capability, we tested 2 participants running simultaneously through remote virtual sessions with our application server and 3 more running at separate times for the same study.

After collecting the logged data we processed the data through the extractor tool. This generates a formatted spreadsheet, and the data can be analyzed, plotted and used as input to other tools for visualization. We collected information on each user's action sequences for a given set of tasks and compared each user's deviation from the error-free best path, described in figures 2 and 3.

3   Results

3.1   Result of the Pilot Study

In order to test the utility and effectiveness of the data capture and begin the analytic phases we ran typical database administrator tasks, for example "create a database index for users". The task of creating an index within a database is trivial in nature (not many steps required), while other included tasks were more complex.

For our first pilot test, we recruited an expert/senior DBA and a new/junior DBA The tool recorded the action sequences, time on tasks, mouse movements and clicks for both the users. The expert user performed the ideal best path through the task and we then measured the deviation from this behavior by the more novice user. For example, on the short task of creating a database instance (not many steps in the ideal, best path), our more novice user did not discover the best path right away. The participant meandered looking for this, including going to external online help to complete the task. This accounted for an extra 26 steps and an extra 5min 58Sec above the expert user to complete the task.

Figure 2 is an example of the automatically logged data output from the extractor tool. We also visualized the deviation using an external tool called Udraw [30]. This deviation from the best path and the time spent in each action sequence

Action Sequence	Action Name	Action type	Action Window(Tar	Target Control Class	Start Time	End Time	Duration
Databases	Expand Item	outline item	Object Explor	WindowsForms10.SysTreeView32.app.0.3	04/20/200	04/20/200	00:06.4
adventureworks2	Expand Item	outline item	Object Explor	WindowsForms10.SysTreeView32.app.0.3	04/20/200	04/20/200	00:04.3
Tables	Left Button Click	outline item	Object Explor	WindowsForms10.SysTreeView32.app.0.3	04/20/200	04/20/200	00:08.0
Sales.Customer	Expand Item	outline item	Object Explor	WindowsForms10.SysTreeView32.app.0.3	04/20/200	04/20/200	00:04.2
Indexes	Expand Item	outline item	Object Explor	WindowsForms10.SysTreeView32.app.0.3	04/20/200	04/20/200	00:02.2
Look for:	Left Button Click	editable text	Index	Edit	04/20/200	04/20/200	00:04.0
Indexes	Right Button Cli	outline item	Object Explor	WindowsForms10.SysTreeView32.app.0.3	04/20/200	04/20/200	00:02.0
New Index...	Left Button Click	menu item	DropDown	WindowsForms10.Window.808.app.0.33c	04/20/200	04/20/200	00:06.0
Help	Left Button Click	push button	New Index	WindowsForms10.ToolbarWindow32.app.	04/20/200	04/20/200	00:12.0
Look for:	Left Button Click	editable text	Index Propert	Edit	04/20/200	04/20/200	00:03.0
Look for:	Left Button Click	editable text	Index	Edit	04/20/200	04/20/200	00:01.7
C2 Audit Mode option	SetEditBoxValu	list item	Index	hx_winclass_vlist	04/20/200	04/20/200	00:00.3
crash recovery [SQL Server]	SendKey	list item	Index	hx_winclass_vlist	04/20/200	04/20/200	00:00.4
CREATE ACTION statement	SendKey	list item	Index	hx_winclass_vlist	04/20/200	04/20/200	00:02.2
CREATE INDEX statement	SendKey	list item	Index	hx_winclass_vlist	04/20/200	04/20/200	00:02.1
CREATE INDEX statement	SendKey	list item	Index	hx_winclass_vlist	04/20/200	04/20/200	00:00.5
Look for:	SendKey	editable text	Index	Edit	04/20/200	04/20/200	00:11.5
Look for:	Left Button Click	editable text	Index	Edit	04/20/200	04/20/200	00:01.3
ScrapReason table [Adventure	SetEditBoxValu	list item	Index	hx_winclass_vlist	04/20/200	04/20/200	00:00.1
ssbdiagnose	SendKey	list item	Index	hx_winclass_vlist	04/20/200	04/20/200	00:00.3
SslPort property	SendKey	list item	Index	hx_winclass_vlist	04/20/200	04/20/200	00:00.6
SslPort property	SendKey	list item	Index	hx_winclass_vlist	04/20/200	04/20/200	00:00.6
Minimize	SendKey	push button	Index	wndclass_desked_gsk	04/20/200	04/20/200	00:33.3
CREATE INDEX (Transact-SQL)	Left Button Click	push button	HttpProtocol.	ToolbarWindow32	04/20/200	04/20/200	00:04.6
Address	Left Button Click	editable text	Running Appli	Edit	04/20/200	04/20/200	00:03.6
Address	Left Button Click	editable text	CREATE INDE	Edit	04/20/200	04/20/200	00:05.6
Address	SetEditBoxValue	editable text	CREATE INDE	Edit	04/20/200	04/20/200	00:00.5
Search for	SendKey	editable text	CREATE INDE	Internet Explorer_Server	04/20/200	04/20/200	00:03.2
SSMS	Left Button Click	editable text	Live Search -	Internet Explorer_Server	04/20/200	04/20/200	00:07.6
Minimize	SendKey	push button	Live Search -	VIEFrame	04/20/200	04/20/200	00:00.4
Index Name	SendKey	editable text	Live Search -	WindowsForms10.EDIT.app.0.33c0d9d	04/20/200	04/20/200	00:10.2
Index Name	Left Button Click	editable text	ssms create i	WindowsForms10.EDIT.app.0.33c0d9d	04/20/200	04/20/200	00:13.9
Add...	Left Button Click	push button	SSMS - Windo	WindowsForms10.BUTTON.app.0.33c0d9	04/20/200	04/20/200	00:02.6
row 0, column 0	Left Button Click	check box	New Index	WindowsForms10.Window.8.app.0.33c0d	04/20/200	04/20/200	00:07.7
OK	SetEditBoxValue	push button	New Index	WindowsForms10.BUTTON.app.0.33c0d9	04/20/200	04/20/200	00:04.8
OK	Left Button Click	push button	New Index	WindowsForms10.BUTTON.app.0.33c0d9	04/20/200	04/20/200	00:02.4
OK	Check	push button	sqlManagerUI	WindowsForms10.BUTTON.app.0.33c0d9	04/20/200	04/20/200	00:03.2
OK	Left Button Click	push button	Select Column	WindowsForms10.BUTTON.app.0.33c0d9	04/20/200	04/20/200	00:02.7

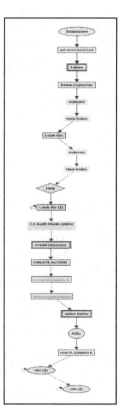

Fig. 2 Best Path Analysis using Expert user vs. Novice user. The rows highlighted in green are the expert behavior and the rows highlighted in pink are the deviation from the expert behavior

Fig. 3 Deviation Visualization

can be useful for researchers and product teams to understand the mental model of real end users.

3.2 Result of Remote Unattended Study Using TCA Logger

We deigned the task of database installation and uninstallation and tested the task using our virtual lab setup. Our participants completed the study across different geographic locations. Since it was our first study using remote mechanism we decided to record the study using Live Meeting [11]. Administering the study helped us to understand latency issues which could be associated with the tool and other network connectivity and performance aspects of the tool. We were fortunate that our participants used varied network connections, from a moderate speed Comcast cable internet connection to very high speed corporate network connections This helped us to understand the reliability of the tool and its capability for use in different network environments.

Before study we had a clear understanding of actual step sequences and count required to complete the tasks (the expected best path). We captured the data for the five participants and analyzed these to see the deviations from best path and the test participants' task-sequence patterns, shown in Figure 1. Looking at the logged data we were able play back the users' action sequences. We found almost zero latency issues for all the cases during our data collection. We were successful in identifying the key approach differences and commonalities across participants. For example, we found participants (1,2,4) used windows authentication mode and participants (3,5) used mixed mode during installation process as fitted with their work environment. Similarly, we found participants (2, 3,4) consulted with the help while participants (1,5) did not.

3.3 Result of a Lab Study Using TCA Logger

Users' self reported their satisfaction scores at the end of each task. The primary aim of the study was to explore how the new SQL Server product interface facilitated the users' task and what if any refinements might be indicated. 5 out of 6 users were able to complete tasks 1 through 5 and their satisfaction increased with each succeeding task (order effect). 6 participants were recruited for a lab- based user study to test a new database report builder interface and, at the same time, test the TCA automated data collection routines. The recruited participants were selected based on a demographics profile of the typical target audience of the product. The participants were given six tasks to complete and were given a maximum of 2 hours. The tasks, among other things, included creating a blocked table with groups and subgroups, and creating a report. We used OvoLogger to video record the participants' action sequence. In addition, the TCA automated data collection tools logged the action sequences of the participants' tasks.

Users' self reported their satisfaction scores at the end of each task. The primary aim of the study was to explore how the new SQL Server product interface

Action Name	ion Window (Targ	Action Type	Action Sequence	arget Control Clas	Start Time	End Time	Duration
Left Button Clic	Text Box Propertie	outline item	Line_Total	WindowsForms10.9	04/29/2008	04/29/2008 19	00:01.2
Left Button Clic	Standard Pane	client	Preview	WindowsForms10.\	04/29/2008	04/29/2008 19	00:01.5
Left Button Clic	Text Box Propertie	client	Design	WindowsForms10.\	04/29/2008	04/29/2008 19	00:01.5
Left Button Clic	Study_Start - Mcr	graphic	Line_Total1	WindowsForms10.\	04/29/2008	04/29/2008 19	00:15.2
Left Button Clic	Desktop	push button	OK	WindowsForms10.8	04/29/2008	04/29/2008 19	00:02.0
Right Button C	Study_Start - Mcr	client	Preview	WindowsForms10.\	04/29/2008	04/29/2008 19	00:01.4
Left Button Clic	Study_Start - Mcr	client	Design	WindowsForms10.\	04/29/2008	04/29/2008 19	00:01.1
Left Button Clic	Study_Start - Mcr	push button	End Task 4	WindowsForms10.8	04/29/2008	04/29/2008 19	00:02.6
Left Button Clic	Study_Start - Mcr	push button	Next	ThunderCommandE	04/29/2008	04/29/2008 19	00:01.0
Left Button Clic	Text Box Propertie	push button	OK	WindowsForms10.8	04/29/2008	04/29/2008 19	00:02.5
Left Button Clic	Text Box Propertie	push button	View	WindowsForms10.\	04/29/2008	04/29/2008 19	00:02.6
Check	Text Box Propertie	client	Preview	WindowsForms10.\	04/29/2008	04/29/2008 19	00:05.4
Left Button Clic	Text Box Propertie	client	Design	WindowsForms10.\	04/29/2008	04/29/2008 19	00:02.3
Right Button C	Study_Start - Mcr	push button	Insert	WindowsForms10.\	04/29/2008	04/29/2008 19	00:02.6
Left Button Clic	Study_Start - Mcr	push button	OK	WindowsForms10.8	04/29/2008	04/29/2008 19	00:01.9
Right Button C	Study_Start - Mcr	client	Preview	WindowsForms10.\	04/29/2008	04/29/2008 19	00:10.0
Left Button Clic	DropDown	client	Design	WindowsForms10.\	04/29/2008	04/29/2008 19	00:02.3
Left Button Clic	Study_Start - Mcr	graphic	Group3	WindowsForms10.\	04/29/2008	04/29/2008 19	00:01.2
....	....	....	....	....			
....	....	....	....	....			

Fig. 4 Small Sample of the TCA Logged data in Design-Preview analysis

facilitated the users' task and what if any refinements might be indicated. 5 out of 6 users were able to complete tasks 1 through 5 and their satisfaction increased with each succeeding task (order effect).

Based on both the logged data and self reported scores captured by TCA on a 7-pt Likert scale, we found that- task six was a difficult task. No user was able to complete this task and users' satisfaction ratings dropped (see figure 5).

The main aim of the study was to assess and validate how the user interface supported the users' tasks. Task 6 required users to create a specific type of report format. It was a hard task for all the participants. It was clear from the TCA data that most users would be best served by a WYSIWYG (what you see is what you get) model. Data showed that participants heavily depended on a "preview" mode

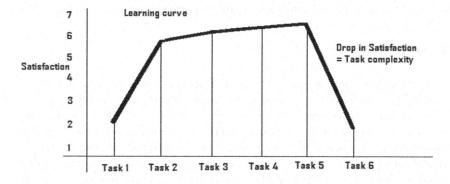

Fig. 5 Satisfaction curve from TCA logged data

to guide and check their work, by switching back and forth between the existing design and preview modes. The nature of the tasks and the supporting user interface and the data showed that increasing satisfaction rate can be labeled as a "learning curve" the users.

While observing the participants activity with the interface, the test administrators felt that most of the users were able to understand and use the preview concept. they noted the frequency of use and, at the end of the study the TCA logged data identified the exact frequency and number of steps incurred by the users' need for and reliance on the preview mode, as the design mode was not WYSIWYG. See table 1 above. We cross referenced these data with the number of clicks between design-preview mode observed from the video analysis. The great many clicks between design and preview modes implied that users were uncertain that what they created in the design mode would achieve their desire format for the report. Every time a user made a change in their report format in the design mode, they wanted to preview it.

Table 1 The number of clicks and time spent per participant switching between Design-Preview modes

Participants	# of clicks between Design-Preview mode	Total time (min : sec)	# of Clicks of Undo	Total time (min : sec)
Participant1	198	37:03	28	4:40
Participant2	203	41:20	20	4:15
Participant3	165	34:15	17	3:48
Participant4	188	43:16	14	3:15
Participant5	137	28:17	16	3:43
Participan6	228	39:08	16	4:01

An average of 37 minutes was spent by the users in going back and forth between design/preview mode in a two hours study. The information was useful in identifying assessing the product design, both in identifying needed design changes (task6) and in validating successful/supportive design tasks 1 through 5.

4 Discussions

The goal of this study was to evaluate the feasibility and value of a user action tracking tool, such as TCA, in conducting remote, unattended usability tests. Our objective was to determine whether we have collectively reached a sufficiently high enough state of the art to support making this a routine and mainstream part of our usability engineering process and, if not, to advance the state further. As shared through the results, we found that the type of data we can now collect through a tool such as TCA can be very effective in analyzing user action sequences collected in remote, unattended studies. At the same time, we also acknowledge that it is critical that the usability engineer who will analyze the data, without having viewed the session directly, has reasonable prior knowledge about the user interface in addition to the user tasks.

This type of data collection works well for remote unattended studies, remote attended studies, and even for in-lab (not remote) studies logged and administered in classic, traditional ways, providing additional data precision that helps teams assess, make, and influence design decisions and business practices. Since these data are directly captured from the users' behavior, we found that product teams tend to consider them less biased and appear to be very interested in these types of data. For example, data from the TCA tool tracking can generate metrics for user researchers and product designers to determine how much usage is required for a user to become proficient, if collected and analyzed over a period of time, using repeated within-subjects remote unattended usability studies (low cost). The longitudinal collection of these data has the potential to analyze intuitiveness and the projected learning curve that will be experienced later in the marketplace based on a planned design.

However, while we have concluded that benchmarking and unattended usability studies are well supported in this way, the usability engineer role, though not necessarily present during the data collection for remote tests, is critical to defining the test protocol for the users' and to the post-test data analysis process. This data collection approach works well for user research that has a set of well-defined task objectives and where the start and end points are well defined and can be recognized.

On the other hand, to use this for more 'real world' field studies, for example in place of a manual diary process where people are typically asked to list and estimate time durations for their activities during a day or week, will require additional work. For example, in addition to the studies already covered to measure this type of capability, we conducted a pilot study with real world database administrators within our IT organization. We collected two small data samples: one with a 30 minute duration and another one with a 2 hour duration.

In this pilot test, the TCA data helped us to identify action sequences, repetitions of actions, and deviations from "best paths". With a short span of 30 minutes and only a few users, we were able to infer and then calculate time slices of each of several critical tasks of interest (for example network load balancing, activity monitoring, backup) and to validate them with the end users. Having this type of transcript may be useful in exploratory field studies, where a usability engineer can sit down with a participant at the end of a data collection period to go over the data together, probing recollections and interpretations in a structured de-briefing interview format. However, it must be recognized that for exploratory studies where the expert path is not known or in a real world environment where the volume of user behavior data is large, where the task objectives and beginning and ending times are ill-defined, rolling up data into accurate granularity, accurate action models to judge the sequences, is a significant challenge.

5 Conclusion and Future Work

Our main interest in designing the TCA tool was to capture and log data to investigate users' behaviors, actions with particular user interface design patterns, through unattended remote usability studies, to expand the participant pool outside a particular geographic region, and to minimize the time & cost in doing so. While we were very interested in the users' behaviors with UI patterns, we were not interested in their data/content – we do not want to compromise security or privacy in any way during these types of remote studies. We therefore made intentional, very explicit and careful commitments to what we would and would not capture & track. In that interest, we do not capture keystrokes, although it could be technically easy to do so during a field test. For example, we capture only users' behaviors with UI objects and field-labels, not open-ended data entry within a field. Key stroke logging has a long history, and we explicitly do not want to capture users' data entry, for example, their logon and password information [5,26]. Therefore, this tool in its present form is not terribly useful for studying scripting languages, though it could be extended to identify particular command language syntax elements to capture. In addition, another refinement we will address in future is the need to track mouse coordinates relative to the screen and the application in use.

This brings us to the question of whether a small sample size of only 6 months is adequate for this kind of analysis. In this paper, we have demonstrated a proof of concept. Our efforts to streamline the data collection using TCA for real world as well as large scale remote unattended usability tests are still tbd. Through theses pilot studies using different conditions, we managed to get "our feet wet", assessing whether we should head into pattern matching, path analytic and statistical modeling/noise reduction exercises in a big way, to infer task objectives, goals and start and end times, from the user behavior patterns themselves. The "state of the art" in the area of applying automated data analytics for the more open-ended type of usage, discovery study is still primitive. It is not easily tractable to field research such as beta and early adopter programs, for assessing 'real world' unguided user behavior, with real task interruptions, multi-tasking, and fits and starts among task objectives, and their starting and ending times. We believe that using this approach to collect real-world user behavior outside of a research mode where task objectives, goals are presented and start and end times are clearly demarcated, will require new work and automation of data analytics.

In summary, we conclude that TCA is a general purpose tool can be used in conducting lab-based studies, both local and remote, and both attended and unattended, and can be used for guided step-by-step pure benchmark studies to assess best performance on the ideal best path in doing tasks, assuming expert, error-free user behavior.

Acknowledgments. As with most endeavors, this paper is not the result of work done solely by the authors. We want to acknowledge key collaborators, including Dawei Huang, George Engelbeck, Kaivalya Hanswadkar , and Mark Stempski for key input and support in refining & testing the TCA tool and remote virtual lab operations, from its early to its present state.

References

1. Badre, A., Paulo, S.J.: Knowledge-Based System for Capturing Human-Computer Interaction Events: CHIME. GVU Technical Report, GIT-GVU-91-21 (1991)
2. BenchmarkSQL: http://sourceforge.net/projects/benchmarksql
3. Brainard, J., Becker, B.: Case Study: E-Commerce Clickstream Visualization. In: Proc. IEEE Proceedings of Information Visualization, pp. 151–155 (2001)
4. Card, S., Moran, T., Newel, A.: The Psychology of Human-Computer Interaction. Lawrence Erlbaum Associates, Mahwah (1983)
5. Ferrer, D.F., Mead, M.: Uncovering the Spy Network: is Spyware watching your library computers? Computers in Libraries 23(5), 16–21 (2003)
6. Greenberg, S., Buxton, B.: Usability evaluation considered harmful (some of the time). In: Proc. CHI 2008, pp. 111–120. ACM Press, New York (2008)
7. Good, M.: The use of logging data in the design of a new text editor. In: Proc. CHI 1985, pp. 93–97. ACM Press, New York (1985)
8. Hartson, H.R., Andre, T.S., Williges, R.C.: Criteria For Evaluating Usability Evaluation Methods. International Journal of Human-Computer Interaction, 373–410 (2001)
9. Hammontree, M., Weiler, P., Nayak, N.: Remote usability testing. Interactions 1(3), 21–25 (1994)
10. Kieras, D.E., Wood, S.D., Kasem, A., Hornof, A.: GLEAN: A Computer-Based Tool for Rapid GOMS Model Usability Evaluation of User Interface Design. In: Proc. UIST 1995, pp. 91–100. ACM Press, New York (1995)
11. LiveMeeting: http://office.microsoft.com/enus/livemeeting/
12. Morae: http://www.techsmith.com/morae.asp
13. MSAA: http://www.microsoft.com/enable/
14. Nielsen, J.: Guerrilla HCI: using discount usability engineering to penetrate the intimidation barrier. Cost-justifying usability, 245–272 (1994)
15. Nielsen, J.: Usability Engineering (1993)
16. Offcial Oracle Benchmark Kits:
   http://www.cisecurity.org/bench_oracle.html
17. Olsen, D.R., Halversen, B.W.: Interface usage measurements in a user interface management system. In: Proc. SIGGRAPH 1988, pp. 102–108. ACM Press, New York (1988)
18. OvoStudio, http://www.ovostudios.com
19. Palmiter, S., Lynch, G., Lewis, S., Stempski, M.: Breaking away from the conventional 'usability lab': the Customer-Centered Design Group at Tektronix, Inc. Behaviour & Information Technology, 128–131 (1994)
20. PolePosiion, http://www.polepos.org/
21. Silverback, http://www.silverbackapp.com
22. SPECjAppServer, http://www.spec.org/order.html
23. Soderston, C.: The Usability Edit: A New Level. Technical Communication 32(1), 16–18 (1985)
24. Soderston, C., Rauch, T.L.: The case for user-centered design. In: Society for Technical Communication annual conference, Denver, Co, USA (1996)
25. Soderston, C.: An Experimental Study of Structure for Online Information. In: 34th International Technical Communication Conference, Society for Technical Communication (1987)

26. Stafford, T.F., Urbaczewski, A.: Spyware: The ghost in the machine. Communications of the Association for Information Systems 14, 291–306 (2004)
27. Stones, C., Sobol, S.: DMASC: A Tool for Visualizing User Paths through a Web Site. In: Proc. International Workshop on Database and Expert Systems Applications 2002 (DEXA 2002), p. 389 (2002)
28. Swingbench: http://www.dominicgiles.com/swingbench.html
29. Tullis, T., Fleischman, S., McNulty, M., Cianchette, C., Bergel, M.: An Empirical Comparison of Lab and Remote Usability Testing of Web Sites. In: Usability Professionals Association Conference (2002)
30. Udraw: http://www.informatik.unibremen.de/uDrawGraph
31. UserFocus,
    http://www.userfocus.co.uk/articles/dataloggingtools.html
32. UserVue, http://www.techsmith.com/uservue.asp
33. Virzi, R.A.: Refining the Test Phase of Usability Evaluation: How Many Subjects Is Enough? Human Factors: The Journal of the Human Factors and Ergonomics Society, 457–468 (1992)
34. Waterson, S., Landay, J.A., Matthews, T.: In the lab and out in the wild: remote web usability testing for mobile devices. In: Proc. CHI 2002, pp. 796–797. ACM Press, New York (2002)
35. Weiler, P.: Software for usability lab: a sampling of current tools. In: Proc. CHI 1993, pp. 57–60. ACM Press, New York (1993)
36. West, R., Lehman, K.R.: Automated Summative Usability Studies: An Empirical Evaluation. In: Proc. CHI 2006, pp. 631–639. ACM Press, New York (2006)

Stepwise Design of BPEL Web Services Compositions:
An Event_B Refinement Based Approach

Idir Ait-Sadoune and Yamine Ait-Ameur

Abstract. Several web services compositions languages and standards are used to describe different applications available over the web. These languages are essentially syntactic ones, their descriptions remain informal and are based on graphical notations. They do not offer any guarantee that the described services achieve the goals they have been designed for. The objective of this paper is twofold. First, it focusses on the formal description, modelling and validation of web services compositions using the Event_B method. Second, it suggest a refinement based method that encodes the BPEL models decompositions. Finally, we show that relevant properties formalized as Event_B properties can be proved. A tool encoding this approach is also available.

1 Introduction

With the development of the web, a huge number of services available on the web have been published. These web services operate in several application domains like concurrent engineering, semantic web or electronic commerce. Moreover, due to the ease of use of the web, the idea of composing these web services to build composite ones defining complex workflows arose.

Nowadays, even if several industrial standards providing specification and/or design XML-oriented languages for web services compositions description, like BPEL [16], CDL [9], BPMN [14] or XPDL [24] have been proposed, the activity of composing web services remains a syntactically based approach. Some of these languages support a partial static semantics checking provided by languages like XML-schema. If these languages are different from the description point of view, they share several concepts in particular the service composition. Among the shared

Idir Ait-Sadoune and Yamine Ait-Ameur
Laboratory of Applied Computer Science (LISI-ENSMA), Téléport 2-1,
avenue Clément Ader, BP 40109, 86961, Futuroscope, France
e-mail: {idir.aitsadoune,yamine}@ensma.fr

Roger Lee (Ed.): SERA 2010, SCI 296, pp. 51–68, 2010.

concepts, we find the notions of *activity* for producing and consuming messages, *attributes* for instance correlation, message decomposition, service location, compensation in case of failure, events and faults handling. These elements are essential to describe services compositions and their corresponding behavior.

Due to the lack of formal semantics of these languages, ambiguous interpretations remain possible and the validation of the compositions is left to the testing and deployment phases. From the business point of view, customers do not trust these services nor rely on them. As a consequence, building correct, safe and trustable web services compositions becomes a major challenge.

Our work proposes to provide a formal semantics for the BPEL language (Business Process Execution Language) and ensure the correctness of web services compositions during the early design phase. We suggest to use formal methods, precisely the Event_B method [2], in order to produce formal models describing web services compositions.

The contribution of this paper is twofold: methodological and technical. From the *methodological point of view*, our objective is not to change the design process mastered by the web services compositions designers, but it is to enrich this process with a formal modelling and verification activity by providing formal models beside each web services composition description. More precisely, our approach consists in deriving an Event_B model from each BPEL web service compositions and encode the decomposition relationship, offered by BPEL, by an Event_B refinement relationship. Therefore, it becomes possible to enrich the Event_B models by the relevant properties in order to check not only the correctness of the web services compositions described in the BPEL designs but also to check the correctness of the decomposition encoded by the sequence of BPEL models linked by the decomposition relationship. Regarding the formal modelling activity, the complexity of the proofs is reduced thanks to refinement.

From a *technical point of view*, we are aware that several work addressed the problem of formal web services verifications [7]. These approaches use model checking as a verification procedure facing the state number explosion problem. Abstraction techniques are used in order to reduce the complexity of the verification. The consequence of this abstraction is the loss of relevant properties modelling and verification like data transformation properties or message exchanges. We claim that, thanks to the Event_B method, we are capable to overcome this limitation by supplying a technique supporting data transformations.

This paper is structured as follows. Next section presents an overview of the state of the art in formal verification of web services compositions. Then sections 3 and 4, describe Event_B method and our approach for transforming any BPEL design into an Event_B model. Static and dynamic BPEL parts are distinguished in this transformation process. Then, we show in section 5 how the refinement offered by Event_B can be used to encode the BPEL decomposition operation. As a result, we obtain an incremental development of both the BPEL design and the Event_B models. Section 6 discusses the properties verification capabilities that result from this approach. Finally, a conclusion and some perspectives are outlined.

2 Formal Validation of Services Compositions: Related Work

Various approaches have been proposed to model and to analyze web services and services compositions, especially formal modelling and verification of BPEL processes. In this section, we overview the formal approaches that have been set up for formal validation of services composition descriptions.

Petri nets were often used to model web service compositions. They have been set up by *Hinz et. al* [15] to encode BPEL processes and to check standard Petri nets properties, and some other properties formalized in CTL logic. *van der Aalst et. al* [1, 25] have defined specific Petri nets called workflow nets to check some properties like termination of a workflow net, and detection of unreachable nodes. Recently, *Lohmann* [17] has completed the BPEL semantics with Petri nets by encoding the new elements appeared in the version 2.0 of BPEL.

Classical transitions systems were also set up to specify web service compositions, especially BPEL processes. We quote the work of *Nakajima* [20, 21] who mapped a BPEL activity part to a finite automaton encoded in Promela with the model checker SPIN to analyze behavioral aspects and to detect potential deadlocks that may occur in the Promela description. In [18, 19], *Marconi et. al* present the approach that translates a BPEL process to a set of state transition systems. These systems are composed in parallel and the resulting parallel composition is annotated with specific web services requirements and is given as input to a set of tools in charge of synthesizing web service compositions expressed in BPEL. FSP (Finite State Process) and the associated tool (LTSA) are used by *Foster et. al* [12, 13] to check if a given BPEL Web Service compositions behaves like a Message Sequence Charts (MSC).

Some work used process algebra for processes and activities formalization. Indeed, *Salaun et. al* [23] show how BPEL processes are mapped to processes expressed by the LOTOS process algebra operations. The same authors in [22] applied their approach to CCS descriptions to model the activities part of a BPEL specification.

Abstract State Machines (ASM) have been used by *Farahbod et. al* [11] to model BPEL workflow descriptions. They take into account exchanged messages and some BPEL real time properties like timeouts. This work has been extended by *Fahland* [10] to model dead-path-elimination.

Other approaches proposed formal models for web service compositions and the BPEL processes. We did not mention them in the above summary due to space limitations. An overview of these approaches can be found in [7].

The whole related work outlined above has two major drawbacks. First, it does not address the description of the static part of BPEL available in the WSDL descriptions. This is due to the abstraction made by the majority of the applied formal methods. Indeed, most of these methods abstract parameters, exchanged messages, preconditions, postconditions and join conditions by Booleans. This abstraction is useful for model checking since it reduces the space explored during verification but it decreases the accuracy of the obtained formal BPEL model. Second, all these proposals translate a BPEL process to another formal model, without offering the

CONTEXT	MACHINE
$c1$	$m1$
SETS	**REFINES**
...	$m2$
CONSTANTS	**SEES**
...	$c1$
AXIOMS	**VARIABLES**
...	...
THEOREMS	**INVARIANTS**
...	...
	THEOREMS
	...
	EVENTS
	...
	END

Fig. 1 The structure of an Event_B development

capability to decompose BPEL precess. So, the decomposition operator offered by BPEL is never addressed by the used formal techniques. The resulting model is often complex to analyze. Thus, the efficiency of the model checking technique often used by existing approaches for checking BPEL processes properties is reduced.

Finally, these methods suffer from the absence of error reporting. Indeed, when errors are detected in the formal model, these approaches do not localize the source of the error on the original BPEL model.

Our work is proof oriented and translates the whole BPEL language and all its constructs into an Event_B model. All the Event_B models presented in this paper have been checked within the RODIN platform[1].

3 Event_B Method

An Event_B model [2] encodes a state transition system where the variables represent the state and the events represent the transitions from one state to another. The refinement capability [3] offered by Event_B allows to decompose a model (thus a transition system) into another transition system with more and more design decisions while moving from an abstract level to a less abstract one. The refinement technique preserves the proved properties and therefore it is not necessary to prove them again in the refined transition system (which is usually more complex). The structure of an Event_B model is given in figure 1.

A **MACHINE** $m1$ is defined by a set of clauses. It may refine another **MACHINE** $m2$. Briefly:

---

[1] The Rodin Platform is an Eclipse-based IDE for Event-B : http://www.event-b.org/platform.html

- **VARIABLES** represents the state variables of the model of the specification.
- **INVARIANTS** describes by first order logic expressions, the properties of the attributes defined in the **VARIABLES** clause. Typing information, functional and safety properties are described in this clause. These properties shall remain true in the whole model.Invariants need to be preserved by events clauses.
- **THEOREMS** defines a set of logical expressions that can be deduced from the invariants. They do not need to be proved for each event like for the invariant.
- **EVENTS** defines all the events that occur in a given model. Each event is characterized by its guard (i.e. a first order logic expression involving variables). An event is fired when its guard evaluates to true. If several guards evaluate to true, only one is fired with a non deterministic choice. The events occurring in an Event_B model affect the state described in **VARIABLES** clause.

An Event_B model may refer to a **CONTEXT** describing a static part where all the relevant properties and hypotheses are defined. A **CONTEXT** *cl* consists of the following clauses:

- **SETS** describes a set of abstract and enumerated types.
- **CONSTANTS** represents the constants used by the model.
- **AXIOMS** describes with first order logic expressions, the properties of the attributes defined in the **CONSTANTS** clause. Types and constraints are described in this clause.
- **THEOREMS** are logical expressions that can be deduced from the axioms.

4   From BPEL Designs to Event_B Models

The formal approach we develop in this paper uses BPEL, one of the most popular web services compositions description language, as a web services design language and Event_B as a formal description technique. We will note indifferently *BPEL design* or *BPEL model* to define the web services composition described in the BPEL language.

4.1   The BPEL Web Services Composition Language

BPEL (Business Process Execution Language [16]) is a standardized language for specifying the behavior of a business process based on interactions between a process and its partners. The interaction with each between partner occurs through Web Service interfaces (*partnerLink*). It defines how multiple service interactions, between these partners, are coordinated to achieve a given goal. Each service offered by a partner is described in a WSDL [8] document through a set of *operations* and handled *messages*. These messages are built from *parts* typed in an XML schema.

A BPEL process uses a set of *variables* to represent the messages exchanged between partners. They also represent the state of the business process. The content of these *messages* is amended by a set of *activities* which represent the process flow. This flow specifies the *operations* to be performed, their ordering, activation

conditions, reactive rules, etc. The BPEL language offers two categories of activities: 1) *basic activities* representing the primitive operations performed by the process. They are defined by *invoke, receive, reply, assign, terminate, wait* and *empty*
activities. 2) *structured activities* obtained by composing primitive activities and/or
other structured activities using the *sequence, if, while* and *repeat Until* composition
operators that model traditional control constructs. Two other composition operators
are defined by the *pick* operator defining a nondeterministic choice based on external events (e.g. message reception or timeout) and the *flow* operator defining the
concurrent execution of nested activities.

In the remained of this paper, we refer to the decomposition of a structured activity by the activities it contains by *the decomposition operation*.

The *invoke, receive, reply* and *assign* activities manipulate and modify the con-
tents of the messages and of the data exchanged between partners. The *if, while* and
repeat Until activities evaluate the content of these messages and data to determine
the behavior of the BPEL process. BPEL also introduces systematic mechanisms
for exceptions and faults handling and also supports the definition of how individual or composite activities, within a unit of work, are compensated in cases of an
exception occurrence or a reversal partner request.

BPEL provides the ability to describe two types of processes: executable and
abstract. An executable process can be implemented in a programming language
and executed by an orchestrator. It describes the desired behavior of the service
orchestration. An abstract process represents a view of a BPEL process. It abstracts
parts or whole of the process and hides implementation details. An abstract process
is not executable.

Finally, an XML representation is associated to the description of every BPEL
process. A graphical representation is available and several graphical editors[2,3]
offer the capability to design BPEL processes. For the rest of the paper, both representations will be used.

4.2 BPEL and the Event_B Semantics

BPEL defines a language for describing the behavior of a business process based
on interactions between the process and its partners. In order to achieve a business
goal, a BPEL process defines how multiple service interactions with these partners
are coordinated together with the logic of this coordination.

An Event_B model encodes a state transitions system where variables represent
the state and events the transitions from one state to another. The refinement capability offered by Event_B makes it possible to introduce more and more design
decisions while moving from an abstract level to a less abstract one. The events of a
model are atomic events and are interleaved defining event traces. The semantics of
an Event_B model is a trace based semantics with interleaving. A system is characterized by the set of licit traces corresponding to the fired events of the model which

---

[2] BPEL Designer : http://www.eclipse.org/bpel/index.php

[3] Orchestra : http://orchestra.ow2.org/xwiki/bin/view/Main/WebHome

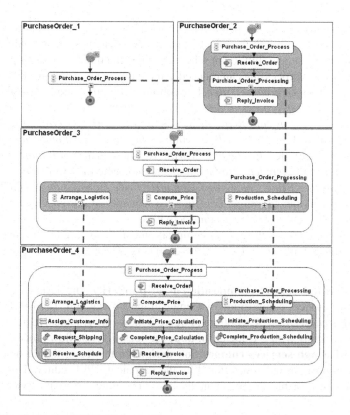

Fig. 2 Graphical editor BPEL decomposition

respects the described properties. The traces define a sequence of states that may be observed by properties.

Our approach for formal modelling of BPEL with Event_B is based on the observation that a BPEL definition is interpreted as a transition system interpreting the processes coordination. A state is represented in both languages by a *variables* element in BPEL and by the VARIABLES clause in Event_B. The various activities of BPEL represent the transitions. They are represented by the events of the EVENTS clause of Event_B.

4.3 A Case Study

The examples used in the following sections are based on the well known case study of the *PurchaseOrder* BPEL specification presented in [16]. This example, issued from electronic commerce, describes a service that processes a purchase order. On receiving the purchase order from a customer, the process initiates three paths concurrently: calculating the final price for the order, selecting a shipper, and scheduling the production and shipment for the order. When some processes are concurrent, they induce control and data dependencies between the three paths. In particular, the shipping price is required to finalize the price calculation, and the shipping date is required for the complete fulfillment schedule. When the three concurrent paths are completed, invoice processing can proceed and the invoice is sent to the customer.

The BPEL model *PurchaseOrder_4* at bottom of Figure 2 shows how the *Purchase_Order_ Process* process is described as a sequence of *Receive_Order*, *Purchase_Order_Processing* and *Reply_Invoice* activities. *Purchase_Order_Processing* is itself decomposed as a flow of *Production_ Scheduling*, *Compute_Price* and *Arrange_Logistics* activities. *Production_Scheduling* is a sequence of *Initiate_ Production_Scheduling* and *Complete_Production_Scheduling* activities. *Compute_Price* is a sequence of *Initiate_Price_Calculation*, *Complete_Price_Calculation* and finally *Receive_Invoice* activities. *Arrange_ Logistics* is a sequence of *Assign_Customer_Info*, *Request_Shipping* and *Receive_Schedule* activities.

When designing the web services composition corresponding to this case study graphical editors tools are set up by the designers. These tools offer two design capabilities.

1. **A one shot web services composition description.** Here the designer produces the whole services composition in a single graphical diagram. The BPEL model named *PurchaseOrder_4* at bottom of figure 2 shows the graphical representation of the case study with a sequence of three services with one flow process.

2. **A stepwise web services composition design.** In this case, the designer uses the decomposition operator offered by the graphical BPEL editor. This operator makes it possible to incrementally introduce more concrete services composing more abstract ones. Figure 2 shows a sequence of three decomposed BPEL models namely *PurchaseOrder_1*, *PurchaseOrder_2*, *PurchaseOrder_3* and *PurchaseOrder_4*.

Nowadays, it is well accepted that formal methods are useful to establish the relevant properties at each step. The presence of refinement during decomposition would help to ensure the correctness of this decomposition. We claim that the Event_B method is a good candidate to bypass these insufficiencies.

4.4   Transforming BPEL Designs to Event_B Models: Horizontal Transformation

This section gives an overview of the horizontal transformation process depicted on figure 3 describing the transformation of any BPEL design to an Event_B model. The proposed approach of [5] is based on the transformation of a BPEL process into an Event_B model in order to check the relevant properties defined in the Event_B models by the services designers. The details of the description of this transformation process and of the rules associated to this transformation process can be found in [5]. For a better understanding of this paper, these rules are briefly recalled below.

This translation process consists of two parts: *static and dynamic*. The BPEL2B tool [6] supports this translation process.

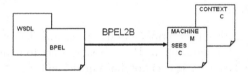

Fig. 3 From BPEL to Event_B: the horizontal transformation

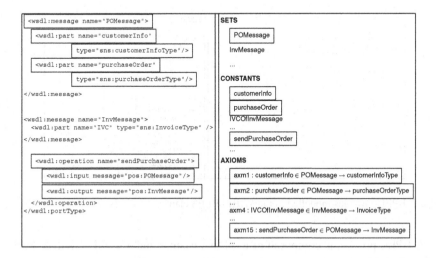

Fig. 4 A message and a port type modelled by an Event_B CONTEXT

4.4.1 Static Part

The first part translates the WSDL definitions that describe the various web services, their data types, messages and port types (the profile of supported operations) into the different data types and functions offered by Event_B. This part is encoded in the **CONTEXT** part of an Event_B model.

A BPEL process references data types, messages and operations of the port types declared in the WSDL document. In the following, the rules translating these elements into a B CONTEXT are inductively defined.

1- The WSDL *message* element is formalized by an abstract set. Each *part* attribute of a *message* is represented by a functional relation corresponding to the template $part \in message \rightarrow type\_part$ from the *message* type to the part *type*. On the case study of figure 4, the application of this rule corresponds to the functions and axioms *axm*1 and *axm*2 declarations.

2- Each *operation* of a *portType* is represented by a functional relation corresponding to the template $operation \in input \rightarrow output$ from the message type of the *input* attribute to the message type of the *output* attribute. On the case study of figure 4, the application of this rule corresponds to the *SendPurchaseOrder* function and axiom *axm*15 declarations.

4.4.2 Dynamic Part

The second part concerns the description of the orchestration process of the activities appearing in a BPEL description. These processes are formalized as B events, each simple activity becomes an event of the B model and each structured or composed activity is translated to a specific events construction. This part is encoded in a **MACHINE** of an Event_B model.

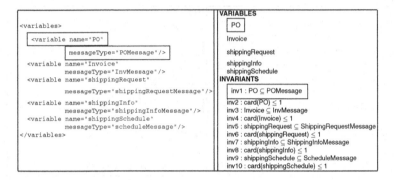

Fig. 5 BPEL variables modelled in the VARIABLES clause.

A BPEL process is composed of a set of variables and a set of activities. Each BPEL variable corresponds to a state variable in the VARIABLES clause, and the activities are encoded by events. This transformation process is inductively defined on the structure of a BPEL process according to the following rules.

3-   The BPEL *variable element* is represented by a variable in the VARIABLES clause in an Event_B **MACHINE**. This variable is typed in the INVARIANTS clause using *messageType* BPEL attribute. The variables and invariants corresponding to the case study are given on figure 5. For example the BPEL variable *PO* is declared and typed.

4-   Each BPEL *simple activity* is represented by a single event in the EVENTS clause of the Event_B **MACHINE**. For example, on figure 6 the *ReceiveInvoice* BPEL atomic activity is encoded by the *ReceiveInvoice* B event. It shows that the operation *SendInvoice* declared in the CONTEXT is referred.

5-   Each BPEL *structured activity* is modelled by an Event_B description which encodes the carried composition operator. Modelling composition operations in Event_B follows the modelling rules formally defined in [4]. Again, on figure 6 the structured activity *Compute_Price* is encoded by a sequence of three atomic activities. A variant *sequ*12 is generated, it decreases from value 3 to the final value 0 in the guard of the *Compute_Price* event.

Notice that all the variables, invariants, constants, typing expressions, events and guards are automatically generated from the information available in BPEL designs by the the BPEL2B tool [6].

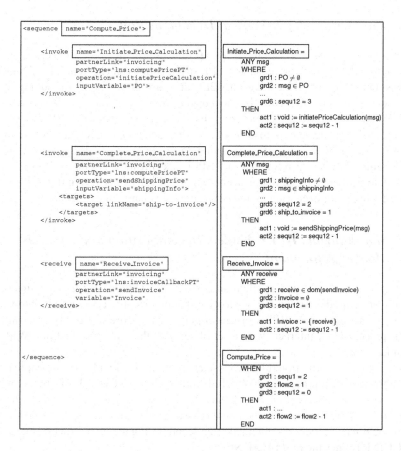

Fig. 6 Atomic and composite (sequence) BPEL activities and their corresponding Event_B representation.

5 From BPEL to Event_B: A Refinement Based Methodology

The previous section addressed the transformation of BPEL models to Event_B models. This transformation process, named *horizontal transformation* produces an Event_B model from any BPEL model whatever is the reached level of the design on the BPEL side. As outlined before, this approach may lead to complex Event_B models with complex proof obligations requiring interactive proofs. It neither takes into account the stepwise decomposition, shown on figure 2, supported by the BPEL descriptions and encoded by the decomposition operator, nor the refinement capability offered by the Event_B method. This section presents a methodology, depicted on figure 7, that exploits Event_B refinement. This section represents the main contribution of this paper.

5.1 The Methodology: Vertical Decomposition

An incremental construction of the BPEL specification using process decomposition operations will help to build reusable BPEL specifications and less complex Event_B models. Our claim is to encode each BPEL decomposition operation (adding new activities) by the refinement development operation of Event_B (adding new events). As depicted on figure 7, we get on the one hand a sequence of abstract BPEL specifications, one being the decomposition of the other, and on the other hand a sequence of proved Event_B refinements corresponding to the abstract BPEL specification. Then a formal stepwise refinement of BPEL specification is obtained. From the methodological point of view, the last obtained BPEL process description is the one that is deployed for execution by the orchestrator in charge of execution.

5.2 Simultaneous BPEL Models Decomposition and Event_B Models Refinement

As shown on figures 2 and 7, when an abstract BPEL process $bpel_i$ contains one or more structured activities, these structured activities are decomposed other subactivities. We get a new BPEL process $bpel_{i+1}$ with more details. Each abstract BPEL process $bpel_i$ interacts with a set of services that are described in a WSDL document $wsdl_i$.

When the transformation rules proposed in sections 4.4.1 and 4.4.2 are applied to each abstract BPEL process and WSDL document, a set of Event_B machines and a set of Event_B contexts are produced. Each $bpel_i$ process is associated to an Event_B MACHINE m_i and each $wsdl_i$ document is associated to an Event_B CONTEXT c_i. When the $bpel_{i+1}$ process is obtained after decomposing the $bpel_i$ process, the m_{i+1} MACHINE refines the m_i MACHINE.

Each MACHINE m_i refers to its CONTEXT c_i. More precisely, when a MACHINE m_{i+1} refines a MACHINE m_i, the CONTEXT c_i referred by the MACHINE m_i is extended by a CONTEXT c_{i+1} by adding definitions of the services introduced by the MACHINE m_{i+1}. Figure 7 illustrates this approach.

5.3 Application to the Case Study

When applied to our case study, the proposed methodology leads to a development of a sequence of 4 Event_B machines, each one refining the previous one. As depicted on the right hand side of figure 7, the Event_B machines $PurchaseOrder\_1$, $PurchaseOrder\_2$, $PurchaseOrder\_3$ and $PurchaseOrder\_4$ define the development of our case study. In order to show the benefits of this approach, we have chosen to comment below the machines obtained after the third refinement.

5.3.1 The Second Refinement

Figure 8 shows part of a machine obtained after refinement. The event *Compute_Price* computes the total amount of the invoice to be produced at the end

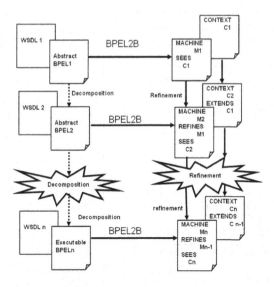

Fig. 7 A refinement based methodology: vertical decomposition.

of shipping. This machine is completed by the relevant resources, represented in grey square boxes, needed to ensure the correct computation of the total amount of the invoice. The invariants *inv*17, *inv*18 and *inv*19 and the action *act*1 are interactively added by the developer. There is no way to generate them from the BPEL code since they are not available.

5.3.2 The Third Refinement

The machine *PurchaseOrder_4* refines the machine *PurchaseOrder_3* as shown on figure 9. Again, the grey square boxes show the information completed by the

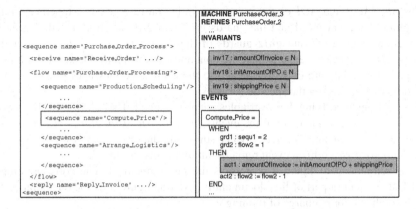

Fig. 8 The second refinement focussing on the Compute_Price event.

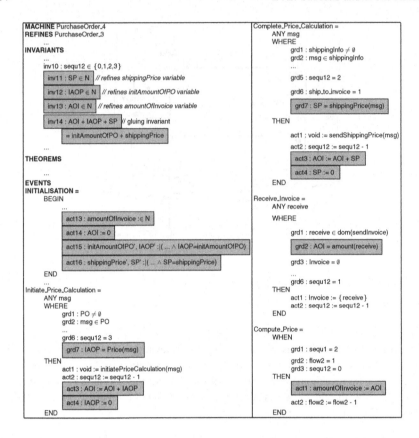

Fig. 9 The third refinement showing the sequence decomposing the Compute_Price event and the gluing invariant.

developer in order to ensure the correctness of the development. The concrete variables *SP*, *IAOP* and *AOI* of the invariants *inv*11, *inv*12 and *inv*13 together with the gluing invariant *inv*14 and the guards *grd*7 and *grd*2 are introduced. The introduction of the *grd*7 and *grd*2 guards of the *Complete_Price_Calculation* and *Receive_Invoice* events, ensures that the concrete variables introduced for defining the gluing invariant are correctly related to the exchanged messages produced by the functions carried by the specific services.

The actions modifying these variables are also defined. They show how the variables evolve. The sequencing of the events *Initiate_Price_Calculation*, *Complete-_Price_Calculation* and *Receive_ Invoice* supplying the result *amountOfInvoice* to the *Compute_Price* refined event remains automatically produced.

The refinement of figure 9 represents the last refinement. The corresponding BPEL document (a part of this document is given on figure 6) is the one deployed by the orchestrator in charge of running it.

6   Validation of Services Properties

When the Event_B models formalizing a BPEL description are obtained, we have seen in the previous section that they may be enriched by the relevant properties that formalize the user requirements and the soundness of the BPEL defined process. In Event_B, these properties are defined in the AXIOMS, INVARIANTS and THEOREMS clauses. Preconditions and guards are added to define the correct event triggering. Our work considers arbitrary sets of values for parameters defining their types. There is no abstraction related to the parameters, preconditions, postconditions nor join operations. The expressions are represented as they appear in BPEL. The proof based approach we propose does not suffer from the growing number of explored states. More precisely, regarding the formal verification of properties, our contribution is summarized in the following points.

- *BPEL type control.* Static properties are described in the CONTEXT of the Event_B model. They concern the description of services, messages and their corresponding types (WSDL part). Event_B ensures a correct description of the types and function composition.
- *Orchestration and services composition.* Dynamic properties are described in the MACHINE of an Event_B model and concern the variables (messages exchanged between services) and the BPEL process behavior (BPEL part). The introduction of variants guarantee the correct services triggering order and message passing.
- *Deadlock freeness.* It is always possible to trigger at least one event. This property is ensured by asserting (in the THEOREMS clause) that the disjunction of all the abstract events guards implies the disjunction of all the concrete events guards. It ensures that if at least one guard of the abstract Event_B model is true, then at least one guard of the refined model is true.
- *No LiveLock.* A decreasing variant is introduced for the purpose of the definition of each refinement corresponding to the encoding of a composition operator. When this variant reaches the value 0, another event may be triggered.
- *Pre-condition for calling a service operation: input message is not empty.* In the orchestration tools, the condition for triggering a BPEL activity is the correct reception of the message used by this activity. Our representation of the call of a service operation takes into account this condition in the events guards corresponding to this activity and an invariant guarantees the existence of this state.
- *Data transformation.* Data properties are expressed in the in the INVARIANTS and AXIOMS clauses. They are checked for each triggered event. This ensures a correct manipulation and transformation of data and messages exchanged between all the participants (partners).
- *Transactional properties.* When modelling fault and compensation handlers by a set of events [5], it becomes possible to model and check properties related to transactional web services.

But, one of the major interests of the proposed methodology is error reporting. Indeed, when an Event_B model cannot be proved due to an erroneous BPEL design and/or decomposition, it is possible to report, on the BPEL code, the occurred error in the concerned activity. This functionality is very helpful for designers that are usually non specialists of formal methods.

Since each BPEL concept is attached to a single Event_B element preserving the same name as in the original BPEL design, it becomes possible to identify, localize and visualize the BPEL concept that corresponds to the B element whose proof obligations cannot be discharged. Notice, that this reporting is only possible for the Event_B parts generated from the BPEL design.

7   Conclusion and Perspectives

This paper presented a fully formalized method for designing correct web services compositions expressed in the BPEL description language. Our contribution addresses both technical and methodological points of view.

From the technical point of view, this approach consists in encoding each BPEL services composition description by an Event_B model in which relevant properties related to deadlock or livelock, data transformation, messages consistence or transactions are modelled. Establishing these properties requires an enrichment of the Event_B models by the relevant information that are not available in the original BPEL description.

From the methodological point of view, our approach suggests to encode the decomposition relationship available on the BPEL modelling side. As a result, the refinement chain of Event_B models follows the decomposition process offered by the BPEL description language. The interest of this approach is double. On the BPEL side it offers a stepwise design approach while it eases the proof activity on the Event_B side since the proof obligations become simpler thanks to the presence of gluing invariants.

Moreover, regarding the approaches developed in the literature, our work covers the whole characteristics of the formal verification of web services compositions. Indeed, the Event_B models we generate support the validation of the properties related to both data (transformation of data) and services (services orchestration).

The BPEL2B tool, presented as an Eclipse PlugIn, encodes the transformation process described in this paper and contributes to the dissemination of formal methods. The details of the formal modelling activities are hidden to the BPEL designer.

Finally, this work opens several perspectives. We consider that two of them need to be quoted. The first one relates to the properties formalization. It is important to formalize more relevant properties. We plan to work more on the transaction based properties. The second one is related to the semantics carried by the services. Indeed, composing in sequence a service that produces distances expressed in centimeters with another one composing services expressed in inches should not be a valid composition. Up to now, our approach is not capable to detect such inconsistencies. Formal knowledge models, ontologies for example, expressed beside Event_B models could be a path to investigate.

References

1. van-der Aalst, W., Mooij, A., Stahl, C., Wolf, K.: Service interaction: Patterns, formalization, and analysis. In: Bernardo, M., Padovani, L., Zavattaro, G. (eds.) SFM 2009. LNCS, vol. 5569, pp. 42–88. Springer, Heidelberg (2009)
2. Abrial, J.R.: Modeling in Event-B: System and Software Engineering, cambridge edn. Cambridge University Press, Cambridge (2010)
3. Abrial, J.R., Hallerstede, S.: Refinement, Decomposition, and Instantiation of Discrete Models: Application to Event-B. Fundamenta Informaticae 77, 1–28 (2007)
4. Aït-Ameur, Y., Baron, M., Kamel, N., Mota, J.-M.: Encoding a process algebra using the Event B method. Software Tools and Technology Transfer 11(3), 239–253 (2009)
5. Ait-Sadoune, I., Ait-Ameur, Y.: A Proof Based Approach for Modelling and Veryfing Web Services Compositions. In: 14th IEEE International Conference on Engineering of Complex Computer Systems, ICECCS 2009, Potsdam, Germany, pp. 1–10 (2009)
6. Ait-Sadoune, I., Ait-Ameur, Y.: From BPEL to Event_B. In: Integration of Model-based Formal Methods and Tools Workshop (IM_FMT 2009), Dusseldorf, Germany (February 2009)
7. van Breugel, F., Koshkina, M.: Models and Verification of BPEL. Draft (2006)
8. Christensen, E., Curbera, F., Meredith, G., Weerawarana, S.: Web Services Description Language (WSDL). W3C Recommendation Ver. 1.1, W3C (2001)
9. Consortium, W.W.W.: Web Services Choreography Description Language (WS-CDL). W3C Recommendation Version 1.0, W3C (2005)
10. Fahland, D.: Complete Abstract Operational Semantics for the Web Service Business Process Execution Language. Tech. rep., Humboldt-Universitat zu Berlin, Institut fur Informatik, Germany (2005)
11. Farahbod, R., Glsser, U., Vajihollahi, M.: A formal semantics for the Business Process Execution Language for Web Services. In: Web Services and Model-Driven Enterprise Information Services (2005)
12. Foster, H.: A Rigorous Approach To Engineering Web Service Compositions. Ph.D. thesis, Imperial College London, University of London (2006)
13. Foster, H., Uchitel, S., Magee, J., Kramer, J.: Model-based Verification of Web Service Composition. In: IEEE International Conference on Automated Software Engineering (2003)
14. Group, O.M.: Business Process Model and Notation (BPMN). OMG Document Number: dtc/2009-08-14 FTF Beta 1 for Version 2.0, Object Manager Group (2009)
15. Hinz, S., Schmidt, K., Stahl, C.: Transforming BPEL to Petri-Nets. In: van der Aalst, W.M.P., Benatallah, B., Casati, F., Curbera, F. (eds.) BPM 2005. LNCS, vol. 3649, pp. 220–235. Springer, Heidelberg (2005)
16. Jordan, D., Evdemon, J.: Web Services Business Process Execution Language (WS-BPEL). Standard Version 2.0, OASIS (2007)
17. Lohmann, N.: A Feature-Complete Petri Net Semantics for WS-BPEL 2.0. In: Web Services and Formal Methods International Workshop, WSFM 2007 (2007)
18. Marconi, A., Pistore, M.: Synthesis and Composition of Web Services. In: Bernardo, M., Padovani, L., Zavattaro, G. (eds.) SFM 2009. LNCS, vol. 5569, pp. 89–157. Springer, Heidelberg (2009)
19. Marconi, A., Pistore, M., Traverso, P.: Specifying Data-Flow Requirements for the Automated Composition of Web Services. In: Fourth IEEE International Conference on Software Engineering and Formal Methods (SEFM 2006), Pune, India (September 2006)
20. Nakajima, S.: Lightweight Formal Analysis of Web Service Flows. Progress in Informatics 2 (2005)

21. Nakajima, S.: Model-Checking Behavioral Specifications of BPEL Applications. In: WLFM 2005 (2005)
22. Salaun, G., Bordeaux, L., Schaerf, M.: Describing and Reasoning on Web Services using Process Algebra. In: IEEE International Conference on Web Service, ICWS 2004 (2004)
23. Salaun, G., Ferrara, A., Chirichiello, A.: Negotiation among web services using LO-TOS/CADP. In: Zhang, L.-J., Jeckle, M. (eds.) ECOWS 2004. LNCS, vol. 3250, pp. 198–212. Springer, Heidelberg (2004)
24. Specification, T.W.M.C.: Process Definition Interface – XML Process Definition Language (XPDL). Document Number WFMC-TC-1025 Version 2.1a, The Workflow Management Coalition (2008)
25. Verbeek, H., van-der Aalst, W.: Analyzing BPEL processes using Petri-Nets. In: Second International Workshop on Application of Petri-Nets to Coordination, Workflow and Business Process Management (2005)

Code Generation for Autonomic Systems with ASSL

Emil Vassev

Abstract. We describe our work on code generation of autonomic systems specified with the Autonomic System Specification Language (ASSL). First, we present
a brief overview of ASSL and then we describe the process of code generation together with features of the generated code in terms of architecture, coding standards,
base classes, and type mapping. Moreover, we demonstrate with samples how the
code generated with ASSL conforms to the ASSL operational semantics. Finally,
we present some experimental results of code generation with ASSL.

1 Introduction

Since 2001, autonomic computing (AC) has emerged as a strategic initiative for
computer science and the IT industry [1]. Many major software vendors, such as
IBM, HP, Sun, and Microsoft, among others, have initiated research programs aim-
ing at the creation of computer systems that exhibit self-management. Although
IBM has released a number of software development tools, there is much to be done
"in making the transition to autonomic culture and it is best to start now" [2].

The Autonomic System Specification Language (ASSL) [3] is a tool dedicated to
AC development that addresses the problem of formal specification and code generation of autonomic systems (ASs) within a framework. ASSL implies a multi-tier
structure for specifying autonomic systems and targets the generation of an operational framework instance from an ASSL specification. In general, ASSL helps to
design and generate an AC wrapper that embeds the components of existing systems
(the so-called managed resource [2, 4]); i.e., it allows a non-intrusive addition of
self-management features to existing systems. Moreover, the framework allows a
top-down development approach to ASs, where the generated framework instance
guides the designers to the required components and their interfaces to the managed

Emil Vassev
Lero-the Irish Software Engineering Research Centre, University College Dublin,
Dublin, Ireland
e-mail: emil.vassev@lero.ie

Roger Lee (Ed.): SERA 2010, SCI 296, pp. 69–85, 2010.
springerlink.com

resource system. In this paper, we present the code-generation mechanism embedded in the ASSL framework. It is important to mention that ASSL generates Java classes forming the skeleton of an AS, which gives the main trend of development but excludes the generation of the implementation of the managed resource [3].

The rest of this paper is organized as follows. In Section 2, we review related work on AS specification and code generation. As a background to the remaining sections, Section 3 provides a brief description of the ASSL framework. Section 4 presents the ASSL code generation mechanism. Section 5 presents some experimental results, and finally, Section 6 presents some concluding remarks and future work.

2   Related Work

A NASA-developed formal approach, named R2D2C (Requirements to Design to Code) is described in [5]. In this approach, system designers may write specifications as scenarios in constrained (domain-specific) natural language, or in a range of other notations (including UML use cases). These scenarios are then used to derive a formal model that fulfills the requirements stated at the outset, and which is subsequently used as a basis for code generation. R2D2C relies on a variety of formal methods to express the formal model under consideration. The latter can be used for various types of analysis and investigation, and as the basis for fully formal implementations as well as for use in automated test case generation. IBM has developed a framework called Policy Management for AC (PMAC) [6] that provides a standard model for the definition of policies and an environment for the development of software objects that hold and evaluate policies. For writing and storing policies, PMAC uses a declarative XML-based language called AC Policy Language (ACPL) [6, 7]. A policy written in ACPL provides an XML specification defining the following elements:

- condition - when a policy is to be applied;
- decision - behavior or desired outcome;
- result - a set of named and typed data values;
- action - invokes an operation;
- configuration profile - unifies result and action;
- business value - the relative priority of a policy;
- scope - the subject of the policy.

The basis of ACPL is the AC Expression Language (ACEL) [6, 7]. ACEL is an XML-based language developed to describe conditions when a policy should be applied to a managed system.

3   ASSL Framework

In general, ASSL considers ASs as composed of autonomic elements (AEs) interacting over interaction protocols. To specify autonomic systems, ASSL uses a multi-tier specification model [3] that is designed to be scalable and to expose a judicious selection and configuration of infrastructure elements and mechanisms needed by an AS. By their virtue, the ASSL tiers are abstractions of different aspects of the AS under consideration, such as `self-management policies`, `communication interfaces`, `execution semantics`, `actions`, etc. There are three major tiers (three major abstraction perspectives), each composed of sub-tiers (cf. Figure 1):

- AS Tier - forms a general and global AS perspective, where we define the general system rules in terms of `service-level objectives` (SLO) and `self-management policies`, `architecture topology`, and global `actions`, `events`, and `metrics` applied in these rules. Note that ASSL express policies with `fluents` (special states) [3].
- AS Interaction Protocol (ASIP) Tier - forms a communication protocol perspective, where we define the means of communication between AEs. The ASIP tier is composed of `channels`, `communication functions`, and `messages`.
- AE Tier - forms a unit-level perspective, where we define interacting sets of individual AEs with their own behavior. This tier is composed of `AE rules` (SLO and self-management policies), an `AE interaction protocol` (AEIP), `AE friends` (a list of AEs forming a circle of trust), `recovery protocols`, special `behavior models` and `outcomes`, `AE actions`, `AE events`, and `AE metrics`.

The ASSL framework comes with a toolset including tools that allow ASSL specifications to be edited and validated. The current validation approach in ASSL is a

```
I. Autonomic System (AS)
   * AS Service-level Objectives
   * AS Self-management Policies
   * AS Architecture
   * AS Actions
   * AS Events
   * AS Metrics
II. AS Interaction Protocol (ASIP)
   * AS Messages
   * AS Communication Channels
   * AS Communication Functions
III. Autonomic Element (AE)
   * AE Service-level Objectives
   * AE Self-management Policies
   * AE Friends
   * AE Interaction Protocol (AEIP)
     - AE Messages
     - AE Communication Channels
     - AE Communication Functions
     - AE Managed Resource Interface
   * AE Recovery Protocol
   * AE Behavior Models
   * AE Outcomes
   * AE Actions
   * AE Events
   * AE Metrics
```

Fig. 1 ASSL Multi-tier
Specification Model

form of consistency checking (handling syntax and consistency errors) performed against a set of semantic definitions. The latter form a theory that aids in the construction of correct AS specifications. Moreover, from any valid specification, ASSL can generate an operational Java application skeleton.

For more details on the ASSL multi-tier specification model and the ASSL framework toolset, please refer to [3]. Note that we are currently working on a model-checking mechanism for ASSL to allow for checking the correctness of both ASSL specifications and generated ASs.

4 ASSL Code Generation

Code generation is the most complex activity taking place in the ASSL framework. In general, it is about mapping validated ASSL specifications to Java classes by applying the ASSL operational semantics [3]. Thus, an operational Java application skeleton is generated for each valid ASSL specification. The former is generated as a fully-operational multithreaded event-driven application with embedded messaging. The code automatically generated by ASSL offers some major advantages over the analytic approach to code implementation. The greatest advantage is that with the generated code we can be certain that our ASSL constructs are properly mapped to correctly generated Java classes. Some other advantages are:

- a logically consistent construction of Java classes is generated;
- the classes in the generated application are linked properly;
- the AS threads are started in the proper order;
- a well-defined messaging system is generated for the ASSL interaction protocols, including event-driven notification, distinct messages, channels, and communication functions;
- a proper type mapping is applied to generate code with appropriate data types;
- appropriate descriptive comments are generated.

4.1 ASSL Code Generation Process

The ASSL framework is a comprehensive development environment that delivers a powerful combination of ASSL notation and ASSL tools [3]. The latter allow specifications written in the ASSL notation be processed to generate Java application skeleton of the same. Figure 2 depicts the process of code generation where the ASSL framework's components collaborate on producing Java classes from ASSL specification structures. Note that here the labeled boxes present different components (e.g., ASSL scanner) in the ASSL toolset and the labeled arrows are data flow arrows. The following elements describe the steps of the ASSL code generation process:

1. The user uses the framework's editor to enter the ASSL specification under consideration and to start the code generation process.

2. The ASSL scanner takes the ASSL specification text and translates the latter into a stream of tokens.
3. The ASSL pre-parser takes the stream of tokens from the scanner and performs special pre-parsing token-consolidation operations on these tokens. In addition, the pre-parser works on the token stream to perform a special declarative evaluation [3] of the ASSL specification and constructs an initial declarative specification tree. The latter is an intermediate graph presenting the ASSL specification in a hierarchical format.
4. The ASSL parser uses the consolidated tokens (cf. ASSL Tokens+ in Figure 2) output from the pre-parser together with the ASSL grammar structures (SLR(1) parsing table expressed with DFAs, First and Follow sets [8]) to parse the ASSL specification. While parsing, the parser prompts the code generator on every shift and reduce parsing operation to generate intermediate code. The parser adds the latter to the nodes of the initial declarative specification tree (the results is Declarative Tree+).
5. The ASSL consistency checker uses the declarative specification tree with embedded intermediate code (Declarative Tree+) to check the consistency of the ASSL specification against special semantic definitions for consistency checking [3].
6. The ASSL Java code generator uses the validated declarative specification tree (with embedded intermediate code) to apply ASSL code generation rules and generates a Java application skeleton.

4.2 AS Architecture for ASSL

By using the ASSL framework, we specify an AS at an abstract formal level. Next, that formal model is translated into a programming model consisting of units and structures that inherit names and features from the ASSL specification. Recall that the framework is responsible for ensuring consistent specification before proceeding to the programming model. The architecture of the generated programming model

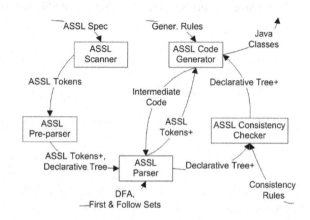

Fig. 2 The Code-generation Process in ASSL

(cf. Figure 3) conforms to the ASSL multi-tier specification model. Every AS is generated with:

- a global AS autonomic manager (implements the AS tier specification) that takes care of the AS-level self-management policies and SLO;
- a communication mechanism (implements the specifications of both ASIP and AEIP tiers; cf. Section 3) that allows AEs to communicate with a set of ASSL messages via ASSL channels;
- a set of AEs (implement the AE tier's specification) where every AE takes care of its own self-management policies and SLO.

Note that both the AS autonomic manager and the AEs incorporate a distinct control loop [2] (cf. Section 4.3.4) and orchestrate the self-management policies of the entire system. However, the AS autonomic manager is a sort of coordinator for the AEs. This coordination goes over AS-level self-management policies, SLO, events, actions, and metrics.

4.2.1 AE Architecture for ASSL

The ASSL generates AEs by applying a special generic architecture model (cf. Figure 4). The latter is considered generic because it is pertaining to any AS successfully generated with the ASSL framework. The initial version of this architecture model was presented at SEAMS 2007 [11].

The AE architecture depicted in Figure 3 is justified by the ASSL multi-tier specification model. Moreover, this is an elaborated version of the IBM blueprint [4] architecture for AEs; i.e., new architecture features are inherited from the ASSL specification model and all the architecture elements have their counterpart in the generated code. Thus, the AE manager forms a `control loop` (cf. Figure 3) over four distinct units namely: (1) `monitor`, (2) `simulator`, (3) `decision maker` and (4) `executor` (the control loop of the AS autonomic manager is generated with the same structure). The four control loop units work together to provide the control loop functionality (cf. Section 4.3.4) by sharing knowledge expressed with instances of the specified for that AE ASSL sub-tiers (e.g., policies, SLO,

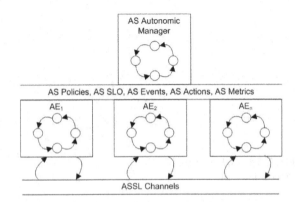

Fig. 3 AS Architecture for ASSL

events, actions). Moreover, the AE uses a set of public channels to communicate ASSL messages with other AEs or environmental entities such as users or external systems and possibly a set of private channels to communicate with trusted AEs (cf. Section 3). The channel controller (cf. Figure 4) controls the AE channels, i.e., it can manage channel settings, close, open or create channels. In addition, it is responsible for sending and receiving ASSL messages over the communication channels. The ASSL framework does not generate a distinct class (component) for the channel controller. Instead its functionality is embedded in the main AE class generated for the AE under consideration. An AE uses a set of `metric sources` to feed with data its metrics. Both metric sources and ME (managed element) effectors operate over the managed resource. Both are specified as special ASSL managed resource interface functions [4] at the AEIP tier (cf. Figure 1). To control both metrics and metric sources, every AE incorporates a metric controller. Similar to the channel controller, the ASSL framework does not generate a distinct class for the metric controller but embeds its functionality in the main AE class.

4.3 Generic Features of the Generated Code

Although every Java application skeleton generated with ASSL comports the specific characteristics of the specified AS, due to the synthetic code-generation approach there are some common generic features of the generated code, which we discuss in this section.

4.3.1 Coding Conventions and Rules

The following elements describe common coding conventions and standards implied by the ASSL framework on the generated code.

- All ASSL tiers/sub-tiers and ASSL named clauses (ASSL structures nested in ASSL sub-tiers) are translated into Java classes.
- All Java classes generated by the ASSL framework have uppercase names.

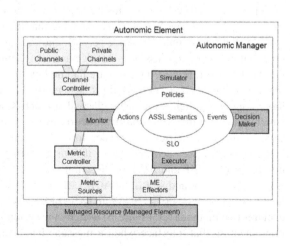

Fig. 4 AE Architecture for ASSL

- All ASSL tier clauses with no specification name and specification variables are translated into methods and data members, those also having uppercase names.
- All generated supplementary data members and methods have names in lower case.
- All base classes generated by ASSL start with an "ASSL" prefix (cf. Section 4.3.2).
- All generated classes are accessed in the code via their qualified names (full package path).
- Every generated class, data member, or method has a descriptive comment in Javadoc style [9].
- All the classes of any generated by ASSL Java application skeleton are located under a commonly-defined hierarchy of Java packages. The root package is "generatedbyassl.as".
- All classes generated by the ASSL framework and mapped to ASSL tiers are implement as singletons [10]; i.e., they define all constructors as private, ensure one instance from a class, and provide a public point of access to it.
- All tier/sub-tier classes, except the main class, implement the Java Serializable interface [9]. The idea behind is that AEs should be able to exchange tier structures at run-time [3].

Please refer to Figures 8 and 9 for samples of some of the coding conventions described above.

4.3.2 Base Classes

The ASSL framework generates base classes (all prefixed with "ASSL") for most of the ASSL tiers/sub-tiers and for some of the ASSL named clauses [3]. These base classes are further used to derive specification-specific classes.

Figure 5 shows a UML class diagram depicting all the base classes generated by the ASSL framework. Here the central class is the base AE class (ASSLAE). Conforming to the structure of the AE Tier (cf. Figure 1), this class declares vectors of tier instances. The latter are instantiated from classes derived from base classes such as ASSLSLO, ASSLPOLICY, etc. Note that there is no base class for the AS tier (this tier has a unique instance per AS, and thus no base class is needed). Also, note that the base AE class does not declare data members for the Friends tier and the AEIP tier (cf. Figure 1). Both are handled by the ASSLAE descendant classes. Moreover, the ASSLAE class defines base methods for controlling the AE control loop, applying policies, etc. This class also inherits the Java Thread class [9], which makes all the AEs running concurrently as threads (processes). All the fluents, events, and metrics are threads too (cf. Figure 5). In addition, among the base classes, we see such that derive the sub-tiers that construct the ASSL interaction protocols ASIP and AEIP (cf. Section 3) - ASSLMESSAGE, ASSLCHANNEL, ASSLFUNCTION, and ASSLMANAGED_ELEMENT. ASSL generates two Java interfaces - ASSLMESSAGECATHCER and ASSLEVENTCATCHER. Whereas the former declares two methods (cf. Figure 5) used by the system to notify instances of classes implementing that interface when a message has been received or sent, the latter declares a method notifying instances of classes implementing that interface when an event has occurred. Note that the base event class (ASSLEVENT) implements both interfaces. For more information on event catchers and all the base classes in general, please refer to [3].

4.3.3 AS Package

Every AS generated by the ASSL framework has an AS Java package, which comprises the main AS class, the AS architecture class, the classes forming the AS control loop, and all the base classes described in Section 4.3.2. In addition, this package nests all the packages comprising Java classes generated for the specified tiers and sub-tiers. Note that the AS class and the AS control loop classes implement the AS autonomic manager (cf. Section 4.2). Figure 6 presents a UML diagram depicting the structure of the AS package. Note that for reasons of clarity, the base classes are not included in this diagram. The central class in Figure 6 is the main AS class. This class defines data members and methods

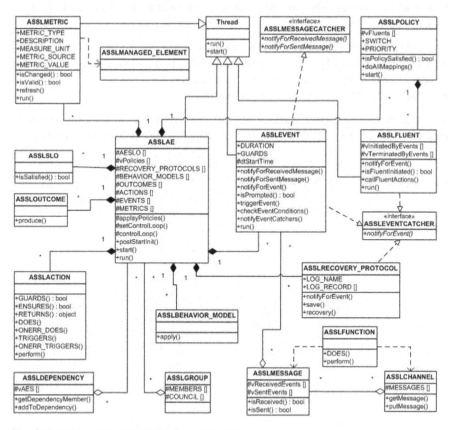

Fig. 5 Base Classes Generated by the ASSL Framework

needed to express the AS tier. Moreover, it extends the Java Thread class and over-
rides the run() and start() methods defined by the latter. In its start()
method, this class starts all the system threads (event threads, metric threads, fluent threads, and AE threads). This class also defines four control-loop data members holding instances of the classes AS_ASSLMONITOR, AS_ASSLANALYZER,
AS_ASSLSIMULATOR, and AS_ASSLEXECUTOR (these inherit the corresponding
base control-loop classes; cf. Section 4.3.4). The AS package also nests a TYPES
package, which comprises classes generated for the custom-defined types in the
ASSL specification.

4.3.4 Control Loop Package

The ASSL framework generates for each AE (specified at the AE tier) and for the AS
Tier a distinct control loop package. The latter comprises four control loop classes
all derived from the base control loop classes. Figure 7 presents a UML diagram
that depicts the structure of the base control loop package. As has been discussed
in Section 4.2.1, the control loop consists of a monitor, a decision maker, a simu-
lator, and an executor. Thus, the base classes generated by the framework for this
package are:

- ASSLMONITOR - base class for AS monitors;
- ASSLANALYZER - base class for all the decision makers in the AS;
- ASSLSIMULATOR - base class for AS simulators;
- ASSLEXECUTOR - base class for AS executors.

Note that these four classes work on SLO (derived from the ASSLSLO class) and
metrics (derived from the ASSLMETRICS class) to discover problems and use

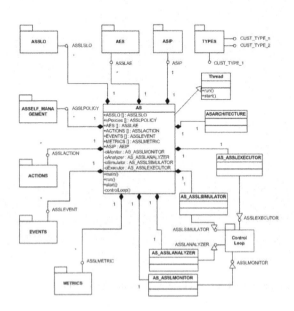

Fig. 6 AS Java Package

actions (derived from the `ASSLACTION` class) to fix those problems. If there is no action set to fix a discovered problem, the control loop executes instance of the `ASSLACTION` class with a generic message notifying about that. The algorithm implemented by the control loop follows the model of behavior exposed by the finite state machines [12]. Thus, we have a finite number of states (`monitoring`, `analyzing`, `simulating`, and `executing`), transitions between those states, and actions (cf. Figure 7). The following elements describe the steps of the control loop algorithm implemented as a finite state machine.

- The finite state machine starts with monitoring by checking whether all the service-level objectives are satisfied and all the metrics are valid (the ASSL operational semantics evaluates both SLO and metrics as Booleans [3]).
- In case there are problematic SLO and/or metrics, the machine transits to analyzing state. In this state, the problems are analyzed by trying to map them to actions that can fix them.
- Next, the machine transits to simulating state. In this state, for all problems still not mapped to actions, the system simulates problem-solving actions in an attempt to find needed ones. Here the control loop applies special AE behavior models [3].
- Finally, the machine transits to executing state, where all the actions determined in both analyzing and simulating states are executed.

The control loop is performed on a regular basis in the `run()` method of both AS and AE classes. Figure 8 shows the code generated for the `run()` method of the main AS class.

4.3.5 Tier Packages

The ASSL framework also generates ASIP/AEIP packages, AE packages, and packages nesting ASSL SLO, ASSL self-management policies, ASSL actions, ASSL

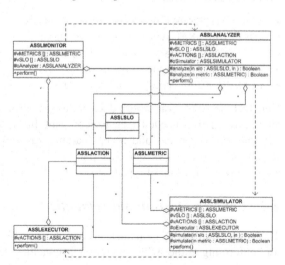

Fig. 7 Base Control Loop Package

Fig. 8 AS class's run()
Method

```
/**
 * Generated by ASSL Framework
 *
 * Runs AS.
 */
public void run (  )
{
    String sMsg = "AS '" + this.getClass().getName() +
    System.out.println( sMsg );
    bStarted = true;
    System.out.println( "*******************************
    //**** runs the control loop
    while ( !bStopAS )
    {
        controlLoop();
        try
        {
            Thread.sleep(100);
        }
        catch ( InterruptedException ex )
        {
            System.err.println( ex.getMessage() );
        }
    }
}
```

events, ASSL metrics, ASSL behavior models, ASSL outcomes, ASSL messages, ASSL channels, and ASSL functions [3].

4.3.6 Type Mapping

In order to generate proper code conforming to the ASSL specification model, the ASSL framework maps the ASSL predefined and custom-specified types to predefined Java types and generated Java type classes. All the ASSL custom-specified types are generated as classes implementing the Java `Serializable` interface [9]. ASSL has predefined types [3], which are mapped by the code generator to Java types as following:

- ASSL INTEGER type - Java `Integer` class;
- ASSL DECIMAL type - Java `Float` class;
- ASSL STRING type - Java `String` class;
- ASSL BOOLEAN type - Java `Boolean` class;
- all ASSL TIME types - Java `Date` class.

Note that the ASSL framework does not use the primitive Java types but their class alternatives. This is needed to ensure compatibility with some generated class methods that return a result and accept parameters of type `Object` (the base class for all the Java classes). For example, the `ASSLACTION` base class implements a `RETURNS()` method of type `Object` (cf. Figure 5).

4.4 Code v/s Operational Semantics

The ASSL framework generates Java application skeletons with run-time behavior conforming to the ASSL operational semantics [3]. In this section we present

Fig. 9 The isPolicySatis-
fied() Method

```
/**
 * Generated by ASSL Framework
 *
 * This method determines whether the policy is satisfied,
 */
public synchronized boolean isPolicySatisfied ( )
{
    boolean bPolicySatisfied = true;
    Enumeration<ASSLFLUENT> eFluents = vFluents.elements();
    ASSLFLUENT currFluent = null;
    while ( eFluents.hasMoreElements() )
    {
        currFluent = eFluents.nextElement();
        //**** A policy is not satisfied if there is at lea
        if ( currFluent.isFluentInitiated() )
        {
            bPolicySatisfied = false;
            break;
        }
    }
    return bPolicySatisfied;
}
```

an example how the generated code for any self-management policy conforms to the ASSL operational semantics rules for policy evaluation. According to the ASSL operational semantics [3], the self-management policies can be evaluated as `Booleans` where if a policy is currently in at least one of its fluents (recall that the ASSL policies are specified with fluents to point at special policy states) it is evaluated as false otherwise it is evaluated as true. Thus, the ASSL framework generates for the `ASSLPOLICY` base class (cf. Figure 5) a method called `isPolicySatisfied()` to evaluate the ASSL self-management policies at run-time. Figure 9 shows the generated implementation of this method. As is shown by Figure 9, the `isPolicySatisfied()` method returns `false` if at least one of the policy fluents is still initiated (`isFluentInitiated()` method), otherwise it returns `true`. This method is used any time when the ASSL framework must evaluate a policy as `Boolean`, e.g., when evaluates `Boolean` expressions over policies. The `isFluentInitiated()` is implemented by the `ASSLFLUENT` base class as a `Boolean` method that evaluates whether a fluent is currently initiated (when a policy enters in a fluent, the latter gets initiated [3]). This method is called any time when the framework evaluates a `Boolean` expression over fluents.

5 Experimental Results

ASSL was successfully used to specify autonomic features and generate proto-typing models for two NASA projects-the Autonomous Nano-Technology Swarm concept mission [13] and the Voyager mission [14]. In this section, we discuss code generation results such as generated code and run-time self-management behavior. The results presented here were obtained by evaluating the successfully generated code for two ASSL specification models for ANTS: self-configuring and self-healing [13].

Fig. 10 ASSL Self-
configuring for ANTS

```
1.  AS ANTS {
2.      ASSELF_MANAGEMENT {
3.          SELF_CONFIGURING {
4.              FLUENT inANTSReconfigurationForNewAsteroid {
5.                  INITIATED_BY { EVENTS.newAsteroidDetected }
6.                  TERMINATED_BY { EVENTS.reconfigurationForNewAsteroidDone }
7.              }
8.              MAPPING { // force ANTS reconfiguration
9.                  CONDITIONS { inANTSReconfigurationForNewAsteroid }
10.                 DO_ACTIONS { ACTIONS.reconfigureANTS }
11.             }
12.         }
13.     } // ASSELF_MANAGEMENT
14.     ACTIONS {
15.         ACTION IMPL reconfigurationForNewAsteroid {
16.             TRIGGERS { EVENTS.reconfigurationForNewAsteroidDone }
17.         }
18.         ACTION reconfigureANTS {
19.             GUARDS { ASSELF_MANAGEMENT.SELF_CONFIGURING.
                                inANTSReconfigurationForNewAsteroid }
20.             ENSURES { EVENTS.reconfigurationForNewAsteroidDone }
21.             DOES { call IMPL ACTIONS.reconfigurationForNewAsteroid }
22.             ONERR_TRIGGERS { EVENTS.reconfigurationForNewAsteroidDenied }
23.         }
24.     } // ACTIONS
25.     EVENTS {
26.         EVENT newAsteroidDetected {
27.             ACTIVATION { CHANGED { AS.METRICS.numberOfAsteroids } }
28.         }
29.         EVENT reconfigurationForNewAsteroidDone { }
30.         EVENT reconfigurationForNewAsteroidDenied { }
31.     }
32.     METRICS {
33.         METRIC numberOfAsteroids {
34.             METRIC_TYPE { RESOURCE }
35.             DESCRIPTION { "the number of detected asteroids during the ANTS lifecycle" }
36.             THRESHOLD_CLASS { DECIMAL [0 ~ ]} // open range: from 0 to ....
37.         }
38.     }
39. } // AS ANTS
```

5.1   Code Generation Statistics

Figure 10 presents the ASSL self-configuring model for ANTS. A detailed description of this model is beyond the scope of this paper. The interested reader is referred
to [13] for more details. For the ASSL self-configuring specification model for
ANTS (cf. Figure 10), the ASSL framework generated 35 Java files distributed into
8 Java packages, which resulted in 3737 lines of code in total. Comparing to the
ASSL code used to specify this model (39 lines of ASSL code; cf. Figure 10) we
get efficiency ratio in terms of lines of code (Java generated code versus ASSL specification code): $3737 / 39 \approx 96$. For the ASSL self-healing model for ANTS [13],
the ASSL framework generated 93 Java files (one per generated class or interface),
which were distributed by the framework into 32 Java packages. The total amount
of generated lines of code was 8159. Comparing to the ASSL self-healing specification model for ANTS, with 293 lines of ASSL code we specified the self-healing
policy at both AS and AE tiers [13]. Therefore, here the efficiency ratio in terms of
lines of code (Java generated code versus ASSL specification code) is: $8159 / 293$
≈ 28. Note that the big difference between the two efficiency rates is coming from
the fact that the ASSL framework generates approximately 2000 lines of code for
the base ASSL Java classes (cf. Section 4.3.2) for each Java application skeleton.
The generation of this code does not directly depend on the ASSL specification under consideration. Based on our experience with ASSL, we consider three levels of
complexity in ASSL specifications:

- low complexity - specifications involving up to 200 lines of ASSL code;
- moderate complexity - specifications involving 201 - 600 lines of ASSL code;
- high complexity - specifications involving over 600 lines of ASSL code.

These efficiency rates show that ASSL provides impressive degree of complexity reduction. This is due to its multi-tier specification model and code generation

mechanism. The efficiency ratio in terms of lines of code (Java generated code versus ASSL specification code) varies between 95-100 for low complex ASSL specifications and 25-30 for ASSL specifications of moderate complexity.

5.2 Generated Code Autonomic Behavior

In this experiment, we experimented with the generated Java application skeletons for the ASSL self-healing specification model for ANTS [13] and for the ASSL specification for the NASA Voyager mission [14]. Our goal was to demonstrate that the ASSL framework generates operational code that is capable of self-managing in respect of the specified ASSL self-management policies. Note that by default, all the ASSL-generated Java application skeletons generate run-time log records that show important state-transition operations going in the generated AS. We used this to trace the behavior of the generated ASs. Although operational the code generated by the ASSL framework is a skeleton, and thus, in order to be fully functional we need to complete the skeleton with the missing parts (generated as empty methods and classes). The presented results were obtained with the generated code only. The evaluation of the log records concluded that the run-time behavior of the generated ASs strictly followed the specified with ASSL self-management policies. It is interesting to mention that in both cases the generated AS followed a common behavior pattern where all system threads are started first. This is shown for the ANTS AS in the following log records.

```
************************************************************
********************* INIT ALL TIERS *********************
************************************************************
******************** START AS THREADS ********************
************************************************************
1)   METRIC 'generatedbyassl.as.aes.ant_ruler.metrics.DISTANCETONEARESTOBJECT': started
2)   EVENT 'generatedbyassl.as.aes.ant_ruler.events.INSTRUMENTLOST': started
3)   EVENT 'generatedbyassl.as.aes.ant_ruler.events.MSGINSTRUMENTBROKENRECEIVED': started
4)   EVENT 'generatedbyassl.as.aes.ant_ruler.events.SPACECRAFTCHECKED': started
5)   EVENT 'generatedbyassl.as.aes.ant_ruler.events.TIMETORECEIVEHEARTBEATMSG': started
6)   EVENT 'generatedbyassl.as.aes.ant_ruler.events.INSTRUMENTOK': started
7)   EVENT 'generatedbyassl.as.aes.ant_ruler.events.MSGHEARTBEATRECEIVED': started
8)   EVENT 'generatedbyassl.as.aes.ant_ruler.events.RECONFIGURATIONDONE': started
9)   EVENT 'generatedbyassl.as.aes.ant_ruler.events.RECONFIGURATIONFAILED': started
10)  EVENT 'generatedbyassl.as.aes.ant_ruler.events.COLLISIONHAPPEN': started
11)  FLUENT 'generatedbyassl.as.aes.ant_ruler.aeself_management.self_healing.INHEARTBEATNOTIFICATION':started
12)  FLUENT 'generatedbyassl.as.aes.ant_ruler.aeself_management.self_healing.INCOLLISION': started
13)  FLUENT 'generatedbyassl.as.aes.ant_ruler.aeself_management.self_healing.INTEAMRECONFIGURATION':started
14)  FLUENT 'generatedbyassl.as.aes.ant_ruler.aeself_management.self_healing.INCHECKINGWORKERINSTRUMENT':started
15)  POLICY 'generatedbyassl.as.aes.ant_ruler.aeself_management.SELF_HEALING': started
16)  AE 'generatedbyassl.as.aes.ANT_RULER': started
************************************************************
17)  METRIC 'generatedbyassl.as.aes.ant_worker.metrics.DISTANCETONEARESTOBJECT': started
18)  EVENT 'generatedbyassl.as.aes.ant_worker.events.ISMSGHEARTBEATSENT': started
19)  EVENT 'generatedbyassl.as.aes.ant_worker.events.INSTRUMENTCHECKED': started
20)  EVENT 'generatedbyassl.as.aes.ant_worker.events.ISMSGINSTRUMENTBROKENSENT': started
21)  EVENT 'generatedbyassl.as.aes.ant_worker.events.COLLISIONHAPPEN': started
22)  EVENT 'generatedbyassl.as.aes.ant_worker.events.INSTRUMENTBROKEN': started
23)  EVENT 'generatedbyassl.as.aes.ant_worker.events.TIMETOSENDHEARTBEATMSG': started
24)  FLUENT 'generatedbyassl.as.aes.ant_worker.aeself_management.self_healing.INHEARTBEATNOTIFICATION':started
25)  FLUENT 'generatedbyassl.as.aes.ant_worker.aeself_management.self_healing.ININSTRUMENTBROKEN':started
26)  FLUENT 'generatedbyassl.as.aes.ant_worker.aeself_management.self_healing.INCOLLISION': started
27)  POLICY 'generatedbyassl.as.aes.ant_worker.aeself_management.SELF_HEALING': started
28)  AE 'generatedbyassl.as.aes.ANT_WORKER': started
************************************************************
29)  EVENT 'generatedbyassl.as.ants.events.SPACECRAFTLOST': started
30)  EVENT 'generatedbyassl.as.ants.events.EARTHNOTIFIED': started
31)  FLUENT 'generatedbyassl.as.ants.assself_management.self_healing.INLOSINGSPACECRAFT': started
32)  POLICY 'generatedbyassl.as.ants.assself_management.SELF_HEALING': started
33)  AS 'generatedbyassl.as.ANTS': started
************************************************************
***************** AS STARTED SUCCESSFULLY *****************
************************************************************
```

Here, records 1 through to 16 show the ANT_RULER AE startup, records 17 through to 28 show the ANT_WORKER AE startup, and records 29 through to 33 show the last startup steps of the ANTS AS. After starting up all the threads, the generated systems run in idle mode until an ASSL event initiating a self-management policy's fluent occurs in the AS. The occurrence of such an event activated the self-management mechanism of the generated AS.

6   Conclusion and Future Work

We have presented a code generation mechanism for generating ASs from their ASSL specification. We presented the generic architecture for both ASs and AEs generated with ASSL. Moreover, we presented the internal structure of the generated code and demonstrated with an example, how the generated code copes with the ASSL operational semantics. In addition, some experimental results demonstrate the efficiency of the ASSL code generator. Due to the log records produced by the generated ASs, we were able to trace their self-management behavior and to conclude that the generated code had followed correctly the specified self-management policies.

Future work is mainly concerned with optimization techniques for the ASSL code generator. In addition, we are currently working on a model checking mechanism for ASSL, which will allow checking safety and liveness properties of the ASSL specifications.

Acknowledgements. This work was supported in part by IRCSET postdoctoral fellowship grant at University College Dublin, Ireland, and by the Science Foundation Ireland grant 03/CE2/I303_1 to Lero (The Irish Software Engineering Research Centre.

References

1. Parashar, M., Hariri, S. (eds.): Autonomic Computing: Concepts, Infrastructure and Applications. CRC Press, Boca Raton (2006)
2. Murch, R.: Autonomic Computing: On Demand Series. IBM Press, Prentice Hall (2004)
3. Vassev, E.: Towards a Framework for Specification and Code Generation of Autonomic Systems, Ph.D. Thesis in the Department of Computer Science and Software Engineering. Concordia University, Montreal, Canada (2008)
4. IBM Corporation. An Architectural Blueprint for Autonomic Computing, White Paper, 4th edn. IBM Corporation (2006)
5. Hinchey, M., Rash, J., Rouff, C.: Requirements to Design to Code: Towards a Fully Formal Approach to Automatic Code Generation, Technical Report TM-2005-212774. NASA Goddard Space Flight Center, Greenbelt, USA (2005)
6. IBM Tivoli. Autonomic Computing Policy Language, Tutorial. IBM Corporation (2005)
7. Agrawal, D., et al.: Autonomic Computing Expressing Language, Tutorial. IBM Corporation (2005)
8. Louden, K.C.: Compiler Construction - Principles and Practice. PWS, Boston (1997)

9. Sun Microsystems. How to Write Doc Comments for the Javadoc Tool. SDN (Sun Developer Network) (2004),
   http://java.sun.com/j2se/javadoc/writingdoccomments/
10. Gamma, E., Helm, R., Johnson, R., Vlissides, J.: Design Patterns: Elements of Reusable Object-Oriented Software. Addison-Wesley, Reading (1995)
11. Vassev, E., Paquet, J.: Towards an Autonomic Element Architecture for ASSL. In: Proceedings of International Workshop on Software Engineering for Adaptive and Self-Managing Systems (SEAMS 2007). IEEE Computer Society, Los Alamitos (2007), doi:10.1109/SEAMS.2007.21
12. Blaha, M., Rumbaugh, J.: Object-Oriented Modeling and Design with UML, 2nd edn. Pearson, Prentice Hall, New Jersey (2005)
13. Vassev, E., Hinchey, M., Paquet, J.: Towards an ASSL Specification Model for NASA Swarm-Based Exploration Missions. In: Proceedings of the 23rd Annual ACM Symposium on Applied Computing (SAC 2008). ACM, New York (2008), doi:10.1145/1363686.1364079
14. Vassev, E., Hinchey, M.: Modeling the Image-processing Behavior of the NASA Voyager Mission with ASSL. In: Proceedings of the Third IEEE International Conference on Space Mission Challenges for Information Technology (SMC-IT 2009). IEEE Computer Society, Los Alamitos (2009), doi:10.1109/SMC-IT.2009.37

A UML Based Deployment and Management Modeling for Cooperative and Distributed Applications

Mohamed Nadhmi Miladi, Fatma Krichen, Mohamed Jmaiel, and Khalil Drira

Abstract. Thanks to the major evolutions in the communication technologies and in order to deal with a continuous increase in systems complexity, current applications have to cooperate to achieve a common goal. Modeling such cooperatives applications should stress regular context evolutions and increasingly users requirements. Therefore, we look for a model based solution suitable to cooperative application that can react in response to several unpredictable changes. Driven by the cooperative application structure, we propose, in this paper, an UML extension named "DM profile" ensuring a high-level description for modeling the deployment and its management in distributed application. The proposed contribution is validated through a "Follow Me" case study and implemented through an Eclipse plug-in.

1 Introduction

Current distributed systems are continuously increasing in size and especially in complexity. Cooperating several software entities, to achieve a common goal, is a key to cope with such complexity. These cooperative applications have to adapt their deployed architectures due to several purposes: improving performance, evolutionary user requirements, context changes, etc.

Mohamed Nadhmi Miladi
University of Sfax, ReDCAD laboratory, ENIS, Box.W 1173, 3038, Sfax, Tunisia
e-mail: `MohamedNadhmi.Miladi@isimsf.rnu.tn`

Fatma Krichen
University of Sfax, ReDCAD laboratory, ENIS, Box.W 1173, 3038, Sfax, Tunisia
e-mail: `Fatma.Krichen@irit.fr`

Mohamed Jmaiel
University of Sfax, ReDCAD laboratory, ENIS, Box.W 1173, 3038, Sfax, Tunisia
e-mail: `mohamed.jmaiel@enis.rnu.tn`

Khalil Drira
CNRS; LAAS; 7 avenue du colonel Roche, F-31077 Toulouse, France
Université de Toulouse; UPS, INSA, INP, ISAE; LAAS; F-31077 Toulouse, France
e-mail: `khalil@laas.fr`

Roger Lee (Ed.): SERA 2010, SCI 296, pp. 87–101, 2010.

A successful adaptation is based on providing architecture deployment models that can be dynamically managed to meet such required purposes. This deployment and management modeling should be, in addition, suitable to the distributed and cooperative features of the application. However, modeling the architecture deployment and its management is often closely coupled to the underlying supported platform. Such a description especially targets the modeling of real deployment units such as artifacts, communication links, and computing units. This requires knowing the context and the underlying deployment platforms before starting the deployment process. For the actual systems where adaptively properties are usually unpredictable, such a deployment and its management modeling remains inappropriate. User requests and the application context are continuously evolving. In addition, the availability of deployment structures such as deployment platform and communication flows are not always guaranteed.

The challenge is to design high-level deployment and management solution that can easily handles diverse deployment infrastructures. This modeling should provide not only platform-independent models, but also a modeling abstraction that handles various architecture deployment approaches applicable for the service-oriented and the component-based architectures. This deployment and management modeling should ensure a best effort adaptation while taking advantage of all available resources whatever their architecture development approach or their underling platform are. Moreover, such modeling should take advantage of the structured organization of the cooperative application.

Basing on the UML standard language, several works propose extensions to handle software architecture deployment such as [9], [12], [3], and [22]. These works follows a structural management reconfiguration on behalf of a deployment management modeling. Other works focus on a high-level modeling that describes the deployment management such as [10], [13], and [17].

The contribution made by this paper merges both: the modeling power of the UML language specification and a high-level modeling. It proposes an UML profile extension providing an abstract model that describes the deployment and its management of distributed software architectures while taking into consideration cooperative architectures specificities. This modeling ensures a best effort solution for the management of an architecture deployment to meet their adaptiveness requirements.

The rest of this paper is organized as follows: Section 2 discuses the related work. In section 3, we present our UML extension profile. Then, in Section 4, we illustrate our extension by a case study Follow Me. Section 5 presents the realization of our profile as a plug-in for eclipse. Finally, section 6 concludes this paper and presents future work directions.

2 Related Work

Many works dynamically manage their software architecture in response to the context evolution requirement. Various management techniques are used. We distinguish the interface management, implementation management, structure management [11], [15]. Other works including [23] merge some of these techniques in order to ensure dynamic management. Despite these research efforts, they remains not well appropriate with the distributed architecture specificities. Works including [13] rely on a deployment management for the context adaptivity requirement of distributed architectures. Modeling dynamic architecture management carries several techniques including ADLs (Architecture Description Languages), formal techniques, graphical techniques.... Several research works focus on standard modeling techniques which are based especially on the UML language and its standard mechanisms. Since our proposed contribution follows a standard modeling techniques and it grants a distributed architecture managing, we focus in this section on researches that handle both: the deployment management and UML extension modeling.

Modeling deployment management through UML extensions can be subdivided into two main issues. The "heavyweight" [21], [22] which set new meta-classes extending UML meta-classes. The other category defines UML profiles while maintaining the UML meta-model. Among these works some researches focus on a context modeling using an UML profile in order to meet the adaptiveness requirements. Some other research efforts including [8], [12], [18] are based on the UML component diagram to achieve a dynamics structural management of architecture applications. However, such works describe a structural management reconfiguration instead of a deployment management modeling.

In order to manage the deployed software architecture, works including [9] are based on a UML profile extending the modeling power of the deployment diagram. Other research efforts addressing the deployment management modeling driven by the architecture type. Works including [20] focus on a service-oriented approach while others including [5] opted for a component-based approach. A third class including [10], [13] and [17] provides a high level description for modeling the deployment and its management.

A final group of research efforts, including [2] merges both: a deployment management modeling based on the UML language and a high-level of abstraction modeling. Our contribution match these efforts while providing a more explicit models for the deployment and its management description especially for cooperative applications. It provides a description that takes advantages of the modeling power of a model based description, a high-level of abstraction modeling and a deployment and management modeling. It is also based on standard direction as well for adopted approach: the MDA approach, as for the used modeling language: the UML language.

3   The Deployment and Management (DM) Profile

This work addresses the deployment and its management modeling for software architectures in general and more specifically for collaborative applications. It proposes models with a high-level of abstraction ensuring a platform independent

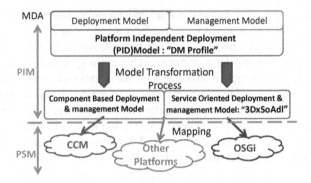

Fig. 1 The multi-level based approach for deploying and managing software architectures

deployment description as well as a best effort management solution for adaptive requirement evolving.

This work is the high-level of a multi-level based approach 1. Based on the MDA approach, the proposed models are transformed towards more specific models through a model transformation process. This process tends to reduce the gap between our abstract models and a more refined model supporting service-oriented [19] or component-based Architectures. These models are mapped towards specific platforms such as CCM [6], OSGi [1], and other platforms as depicted in Figure 1.

3.1   The Deployment Model

The architecture deployment modeling is based on two major ideas. The first idea highlights a deployment model without any prior description of the real deployment architecture entities. This modeling is based on the Unified Modeling Language. The architecture deployment modeling in UML2 is ensured mainly through the deployment diagram. Meta-models of this diagram, which are related to the L2 level of the MOF approach, provide the basic concept for modeling deployment architectures.

In UML2, the deployment modeling process is closely coupled with the real physical architectural entities. Most provided meta-models and versioning update efforts on this diagram follow a modeling vision that focus on the description of physical architecture entities such as "device", "execution environment", "artifact".... Thus, the deployment diagram is associated with a PSM level. In the proposed deployment model, we extend the modeling power of UML deployment diagram to enable a PIM level deployment modeling. This extension ensures a high-level description to model an architectural deployment process. Two new stereotypes are achieved to overtake such deployment modeling:

≪**Deployed Entity**≫ **stereotype:** it ensures the modeling of the functional aspect of the deployment architecture. "Deployed Entity" models each software entity that establishes a functional contract through its provided and required interfaces and can be deployed in a container. Such modeling enhances the description power of UML component model in order to cover several software entities including components, services, service-components [4]... Therefore, it describes the basic software unit in a high deployment level. The defined stereotype extends the component meta-model, as depicted in Figure 2, establishing a more abstract semantic in a deployment context.

≪**Logical Connection**≫ **stereotype:** it ensures the connections modeling between deployed entities of the architecture. The described connections span several connection types ensuring the communication between two "deployed entities". The "Logical Connection" stereotype enhances the connector semantics enabling to span physical connections such wired, wireless or satellites connections, software connections, or connections that express functional dependencies between two "Deployed Entities". This stereotype extends the connector meta-model of UML as depicted in Figure 2.

Modeling software architecture deployment requires the description of, first, the suitable/available deployment containers, second, the deployable software entities such as components and services in their related nodes and, third, the connection links within these entities. This modeling, although it's higher-level description, remains especially focused on concepts reduced to their location typically modeled through the node concept.

Our second basic idea is to extend the container deployment scope using some virtual structures which have a more enhanced semantics than the traditional node concept. These structures ensure a more flexibility and better organization in the deployment process. Each entity will be deployed in virtual structures meeting the

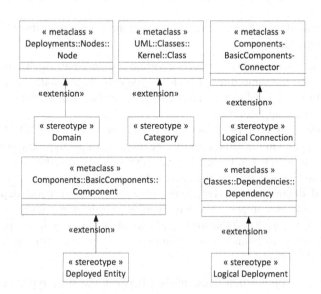

Fig. 2 "DM profile" stereotypes for the deployment modeling

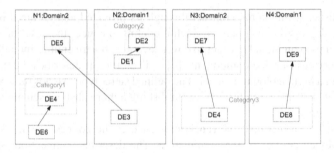

Fig. 3 A description of the cooperative deployment structure

context requirements and accustoming to the available resources. In addition, structuring software entities under some virtual structures emphasize the cooperative aspect of an application and match its cooperative guard. Typically, in these applications, deployed entities are virtually underlying other entities to cooperate in the achievement of the same goal. In a cooperative application, several actors should be established. Each one plays a specific role in the cooperation process. For instance, in a cooperative document edition, various roles are identified such as writer, reader, manager... This carries a vertical vision, as depicted in Figure 3, of the cooperative deployment structure. On the other hand, actors cooperate in the establishment of complex activities to meet a common goal through several activity categories. For example, in a cooperative document edition, we can identify various categories such writing, document correction, review.... This carries a horizontal vision, as depicted in Figure 3, of the cooperative deployment structure. Moreover, it achieves a better management of the deployed architecture. This will be more detailed in the next section.

This structured modeling is achieved through the definition of three new stereotypes on our profile:

≪**Domain**≫ **stereotype:** it models a virtual structure owning "Deployed Entities" that ensure the same role. This stereotype extends the node meta-model, as depicted in Figure 2, establishing a more specific semantics in the deployment context of cooperative application.

≪**Category**≫ **stereotype:** it models a virtual structure owning "Deployed Entities" that cooperate in the same activity category. This stereotype extends the class meta-model, as depicted in Figure 2, establishing a more specific classification of the "Deployed Entities" upon cooperative activities process.

≪**Logical deployment**≫ **stereotype:** it expresses all available dependencies within the profile stereotypes. This stereotype extends the dependency meta-model, as depicted in Figure 2, establishing deployment dependency in cooperative application. Restrictions bound to the definition of this profile depicted in Figure 4 presented in the following:

- A relation of composition is established between the stereotype "Logical Deployment" and the stereotype "Domain", while a deployed architecture includes several "Domain";

- A relation of aggregation is established between the stereotype "Domain" and the stereotype "Category" (respectively "Deployed Entity"), while the same "Category" (respectively "Deployed Entity") can belong to several "Domains";
- A relation of aggregation is established between the stereotype "Category" and the stereotype "Deployed Entity", while the same "deployed entity" can to be deployed in several categories;
- The relation "XOR" between the two previous aggregation relations ensuring that a "Deployed entity" belongs to a "Domain" as well as to a "Category" but not at the same time;
- An association between the stereotype "Deployed Entity" and the stereotype "Logical Connection", while two or more "Deployed entities" can be connected through several "logical connections".

3.2 The Management Model

In order to take over an evolving context and unexpected events or requirements, a dynamic architecture deployment should be modeled. The challenge is to provide notations and mechanisms to cope with the current architecture properties such as large scale deployment while being able to target the appropriate entities to manage. The idea is to describe specific redeployment rules. A redeployment rule describes what's transformation should be achieved on the fly upon the actual deployment architecture to meet the requirement adaptations.

In a large-scale deployment, in order to meet unexpected events or requirements, several redeployment rules should be achieved. In fact, an occurrence of a single

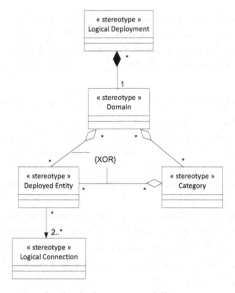

Fig. 4 'DM profile" structure for the deployment modeling

event may lead to achieve a set of redeployment rules. In a cooperative context, such redeployment rules may affect all "deployed entities" without expecting their task in a cooperative process. These rules should target some specific "Deployed Entities" establishing a cooperative task. Describing all these redeployment rules requires both a meticulous and an excessive modeling time and efforts. This modeling solution became especially limited in applications that require a quick and specific management description. The idea is to propose a more expanded redeployment rules. The deployment scopes of such rules are no longer limited to a single "deployed entity" but they handle a set of "deployment entities". The execution of a single expanded deployment rule induces the execution of several underlying deployment rules. Establishing these expanded rules is based on the deployment structures proposed in the previous section including "Domain" and "Category". This enables a more target deployment management driven by the cooperative aspects of the application. Describing these expanded redeployment rules is based on a multi-formalism management approach.

This management approach combines the power of two formalisms. First, it is based on the theoretical efforts achieved on grammar productions techniques such as graph DPO, Δ, Y [16], [12]. We are based especially on the DPO technique. DPO is a richer structure for grammar productions. These productions are specified with a triplet $<L;K;R>$. The application of this production is achieved through the removal of the graph corresponding to the occurrence of Del=$(L\backslash K)$ and the insertion of a copy of the graph Add=$(R\backslash K)$. Indeed, DPO technique, describes each uplet graph as an autonomic entity. There is no exclusion as the Y and/or Δ techniques. This is very useful when, for instance, we should express the relationship of two elements sharing the same container; one should be preserved and the other one should be added.

Second, the proposed management approach is based on the high expressive power of UML language and its standard notations. Based on a graphical notation, our profile provides a clear solution modeling for dynamic deployment management. This solution is achieved through four new stereotypes as depicted in Figure 5:

≪**Rule**≫ **stereotype:** It models an expanded redeployment rule that its description is ensured through the three following stereotypes: ≪**L**≫, ≪**K**≫, and ≪**R**≫ stereotypes. ≪**L**≫ stereotype presents the initial sub-architecture from the system where the redeployment rule can be applied. ≪**K**≫ stereotype presents the sub-architecture to preserve from the sub-architecture stereotyped by ≪**L**≫ stereotype. ≪**R**≫ stereotype present the sub-architecture after the execution of the rule. Otherwise, ≪**L**≫\≪**K**≫ (respectively ≪**K**≫, ≪**R**≫\≪**K**≫) models the deployment structures ("Domain", "Category") and their owning "Deployed Entities" to be deleted (respectively preserved, added) in the rule execution. In other words, after the execution of a redeployment rule the current deployed architecture shifts by adding ≪**R**≫\≪**K**≫ sub-architecture, keeping ≪**K**≫ sub-architecture and deleting ≪**L**≫\≪**K**≫ sub-architecture.

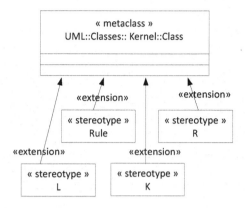

Fig. 5 "DM profile" stereotypes for the management modeling

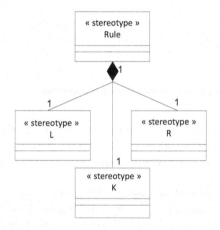

Fig. 6 "DM profile" structure for the management modeling

A restriction, depicted in Figure 6, bound to the definition of this profile. In fact, a relation of composition is established between the stereotype "Rule" and the stereotype "L" ("K" and "R"), because the same sub-architecture "L" ("K" or "R") can belong to only one reconfiguration rule.

4 Case Study: Follow Me

In this section, we present a case study called "Follow Me" for illustrating our profiles. The "Follow Me" case study, which is similar to the one presented in [14], is an adaptive application reacting on the context change. It is an audio application whose audio flow follows the listener movement among many rooms. Each room has some predefined sinks (a player and its speakers). The "Follow Me" architecture ensures the context adaptation thanks to its deployment management

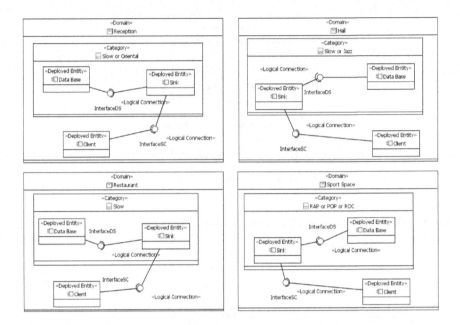

Fig. 7 "Follow Me" high-level deployed structured modeling

ability. In fact, "Follow Me" architecture depends on the available system listeners. If there is no person present in a given room, sinks stop the audio flow. In addition, if a listener moves from a room to another, audio flow follows him/her in a continuous manner.

As an example, we present the case of hotel that has a set of rooms with different roles: reception, hall, restaurant. . . . Each room has a data base which contains songs and players which play music. Rooms sharing the same role play the same music category. Besides, each room can provides to the maximum three players, and each player can serve to the maximum fifteen clients in order to offer better audio flow quality.

In order to describe our case, we need to have some virtual structures which have a more enhanced semantics than the traditional node concept. Thus, we defined the two semantics domains and categories. Additionally, this system must react with the change of context. For that and in order to handle the cooperative aspect, we describe this architecture with a high-level of description: deployed entities interconnect with logical connectors. The following section describe with details our case study "Follow Me" of a hotel.

4.1 Structural Architecture Description

In this section, we propose a modeling process that guide the architecture designer in order to describe a high-level modeling of the deployed structured architecture.

First, we begin by identifying the various architecture types (domains, categories and deployed entities), then we associate categories to domains and deployed entities to domains and categories.

Figure 7 describes deployed structured architecture of the "Follow Me" case study, and in particular the case of hotel, through our realized plug-in and profile.

Identify architecture types:

- Domains: Reception, Hall, Restaurant, Sport Space
- Categories: Slow, Oriental, Jazz, RAP, POP, ROC
- Deployed entities: Sink, Client, Data Base

  - Sink: It provide audio flow
  - Client: He use audio service
  - Data Base: this is an audio data base

- Logical Connections

  - Connection between Data Base and Sink
  - Connection between Sink and Client

Associate categories to domains:

- Reception = Slow, Oriental
- Hall = Slow, Jazz
- Restaurant = Slow
- Sport Space = RAP, POP, ROC

Associate deploymed entities to domains and categories:

- Deployed entities Sink and Data Base belong to the defined categories
- Deployed entities Client belong to defined domains

From this high-level description of the "Follow Me" architecture, we can define several architectural instances of hotels.

Besides, this high level description enables us to specify the different reconfiguration rules of our case study. In the following, we present an example of reconfiguration rules. Such rules model some of elementary redeployment actions that can be applied on deployed structured architecture instance.

4.2 Reconfiguration architecture description

In order to highlight the dynamic management aspect, we consider the case of the arrival of client number sixteen in the reception, a duplication of the deployed entity "Sink" instance should be achieved to serve the new client. Figure 8 models the duplication of "sink1" instance (regarding the deployed entity "Sink") in the category "Slow" in the instance "reception1" regarding the "Reception" domain :

- Add new instance sink2 of deployed entity Sink identical to instance sink1 (same state)

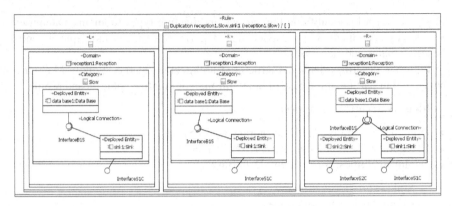

Fig. 8 High-level Rule for the duplication of "Sink" instance

- Add connection between the new instance of deployed entity Sink and the instance data base1 of deployed entity Data Base

5 Eclipse Plug-in Extension

In this section, we provide an UML graphical editor as a plug-in in Eclipse that implements the proposed "DM profile". It ensures a technical issue to model a structural deployment and its management in cooperative distributed architectures. Implementing the "DM Plug-in" is directed by the "UML2Tools" [7] project. This project aims at providing a graphical solution for modeling UML diagrams with respect to their latest version. More specifically, the "DM plug-in" is implemented using the "UML2Tools" deployment diagram. In addition, developing the proposed plug-in is based on several frameworks: the GMF framework (Graphical Modeling Framework) for graphical editor generation, The EMF Framework (Eclipse Modeling Framework) for meta-model construction, and the GEF Framework (Graphical Editing Framework) for graphical drawings.

The Developed "DM Plug-in" did not guarantee only a graphical modeling but also it ensures the mapping of the achieved models towards the XML language as depicted in figure 9. The resulted XML files are generated with respect to the XMI (XML Metadata Interchange) standard recommended by the OMG group. Baring in mind, all XMI files are automatically validated through XML schemas (integrated in the Plug-in). Generating XML files is a fundamental step in a refinement process which starts with already designed models towards a more platform specific description.

Figure 9 shows the different stereotypes defined in our profile and applied in a simple example. For each graphical modelling, an XMI file is generated which contains with details all used UML models and stereotypes.

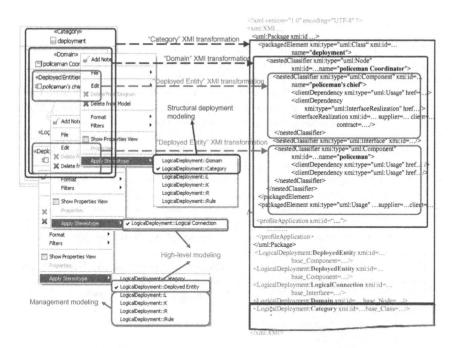

Fig. 9 Mapping of the "DM profile" models towards XML language

6 Conclusion

In this paper, we propose an UML profile, named "DM profile" for the deployment and its management in cooperative and distributed architectures. Driven by the cooperative system structure, the proposed profile ensures: first, a high level description for a deployment modeling decoupled with related platforms and architecture style specificities. Second, an explicit model for managing the deployed architecture based on graph transformation theories. Third, a suitable solution for a large scale deployment of cooperative architectures. The proposed solution is illustrated through a follow me example and implemented through an Eclipse plug-in.

The modeling solution depicted, here, ensures a generic and a platform independent modeling according to the MDA approach. In our future works, we will focus on the model transformation process that fit our multi-level based approach introduced in the top of "The deployment and management (DM) profile" section. This approach seeks to refine the "DM profile" models towards a deployment and management description [19] which can easily mapped to a specific platform such as OSGi. Such model transformation process, is driven by a set of implemented algorithms that translates our high-level models to a set of elementary redeployment actions more suitable with specific platform management.

References

1. Alliance, O.: Sgi service platform core specifcation the osgi alliance (2007), `http://www.osgi.org/Download/Release4V41` (Release 4, version 4.1)
2. Almeida, J.P.A., van Sinderen, M., Pires, L.F., Wegdam, M.: Platform-independent dynamic reconfiguration of distributed applications. In: Proceedings of the 10th IEEE International Workshop on Future Trends of Distributed Computing Systems (FTDCS 2004), pp. 286–291. IEEE Computer Society, Los Alamitos (2004)
3. Ayed, D., Berbers, Y.: UML profile for the design of a platform-independent context-aware applications. In: Proceedings of the 1st workshop on MOdel Driven Development for Middleware (MODDM 2006), pp. 1–5. ACM, New York (2006)
4. Curbera, F.: Component contracts in service-oriented architectures. Computer 40(11), 74–80 (2007), `http://dx.doi.org/10.1109/MC.2007.376`
5. Dearle, A., Kirby, G.N.C., McCarthy, A.J.: A Framework for Constraint-Based Deployment and Autonomic Management of Distributed Applications. In: Proceedings of the 1st International Conference on Autonomic Computing (ICAC 2004), pp. 300–301. IEEE Computer Society, Los Alamitos (2004)
6. Deng, G., Balasubramanian, J., Otte, W., Schmidt, D.C., Gokhale, A.: Dance: A qos-enabled component deployment and configuration engine. In: Dearle, A., Eisenbach, S. (eds.) CD 2005. LNCS, vol. 3798, pp. 67–82. Springer, Heidelberg (2005)
7. Foundation, T.E.: UML2 Tools, `http://www.eclipse.org/modeling/mdt/downloads/project=uml2tools`
8. Göbel, S.: An MDA Approach for Adaptable Components. In: Hartman, A., Kreische, D. (eds.) ECMDA-FA 2005. LNCS, vol. 3748, pp. 74–87. Springer, Heidelberg (2005)
9. Grassi, V., Mirandola, R., Sabetta, A.: A UML Profile to Model Mobile Systems. In: Baar, T., Strohmeier, A., Moreira, A., Mellor, S.J. (eds.) UML 2004. LNCS, vol. 3273, pp. 128–142. Springer, Heidelberg (2004)
10. Grassi, V., Mirandola, R., Sabetta, A.: A model-driven approach to performability analysis of dynamically reconfigurable component-based systems. In: Proceedings of the 6th international workshop on Software and performance, pp. 103–114. ACM, New York (2007)
11. Han, T., Chen, T., Lu, J.: Structure Analysis for Dynamic Software Architecture. In: Proceedings of the 6th ACIS International Conference on Software Engineering, Artificial Intelligence, Networking and Parallel/Distributed Computing (SNPD 2005), p. 338. IEEE Computer Society, Towson (2005)
12. Kacem, M.H., Miladi, M.N., Jmaiel, M., Kacem, A.H., Drira, K.: Towards a UML profile for the description of dynamic software architectures. In: Component-Oriented Enterprise Applications, Proceedings of the Conference on Component-Oriented Enterprise Applications (COEA 2005), LNI, pp. 25–39. GI (2005)
13. Ketfi, A., Belkhatir, N.: Model-driven framework for dynamic deployment and reconfiguration of component-based software systems. In: Proceedings of the 2005 symposia on Metainformatics (MIS 2005), p. 8. ACM, New York (2005)
14. Kirk, R., Newmarch, J.: A Location-aware, Service-based Audio System. In: Proceedings of the Second IEEE Consumer Communication and Networking Conference (CCNC 2005). IEEE Computer Society, Los Alamitos (2005)
15. Letaifa, A.B., Choukair, Z., Tabbane, S.: Dynamic Reconfiguration of Telecom Services Architectures According to Mobility and Traffic Models. In: Proceedings of the 18th International Conference on Advanced Information Networking and Applications (AINA 2004), pp. 447–450. IEEE Computer Society, Fukuoka (2004)

16. Loulou, I., Kacem, A.H., Jmaiel, M., Drira, K.: Towards a Unified Graph-Based Framework for Dynamic Component-Based Architectures Description in Z. In: Proceedings of the The IEEE/ACS International Conference on Pervasive Services (ICPS 2004), pp. 227–234. IEEE, Los Alamitos (2004)

17. Mikic-Rakic, M., Malek, S., Beckman, N., Medvidovic, N.: A Tailorable Environment for Assessing the Quality of Deployment Architectures in Highly Distributed Settings. In: Emmerich, W., Wolf, A.L. (eds.) CD 2004. LNCS, vol. 3083, pp. 1–17. Springer, Heidelberg (2004)

18. Miladi, M.N., Kacem, M.H., Jmaiel, M.: A UML profile and a FUJABA plugin for modelling dynamic software architectures. In: MoDSE 2007: Workshop on Model-Driven Software Evolution, March 20-23. IEEE - CSMR, Amsterdam (2007)

19. Miladi, M.N., Krichen, I., Jmaiel, M., Drira, K.: An xADL extension for managing dynamic deployment in distributed service oriented architectures. In: Proceedings of the 3rd International Conference on Fundamentals of Software Engineering (FSEN), pp. 439–446. Springer, Kish Island (2009)

20. Moo-Mena, F., Drira, K.: Reconfiguration of Web Services Architectures: A model-based approach. In: Proceedings of the 12th IEEE Symposium on Computers and Communications (ISCC 2007), pp. 357–362. IEEE Computer Society, Los Alamitos (2007)

21. Pérez-Martínez, J.E.: Heavyweight extensions to the UML v metamodel to describe the C3 architectural style. SIGSOFT Softw. Eng. Notes 28(3), 5 (2003)

22. Poggi, A., Rimassa, G., Turci, P., Odell, J., Mouratidis, H., Manson, G.A.: Modeling Deployment and Mobility Issues in Multiagent Systems Using AUML. In: Giorgini, P., Müller, J.P., Odell, J.J. (eds.) AOSE 2003. LNCS, vol. 2935, pp. 69–84. Springer, Heidelberg (2004)

23. Walsh, D., Bordeleau, F., Selic, B.: A Domain Model for Dynamic System Reconfiguration. In: Briand, L.C., Williams, C. (eds.) MoDELS 2005. LNCS, vol. 3713, pp. 553–567. Springer, Heidelberg (2005)

Development of Mobile Location-Based Systems with Component

Haeng-Kon Kim and Roger Y. Lee

Summary. The current mobile embedded software development technologies are only rooted on specific platforms means platform dependent. so a real designing engineering methodology is needed. seeks to define, construct, validate and deploy a new model-based methodology and an interoperable toolset for real-time embedded systems development.

The proxy driving service is having a boom recently, which a proxy driver on behalf of a drunken one has a car to the destination at night. The call center selects the nearest proxy driver based on the distance from the customer and sends customer's information to the designated one. A proxy driver usually speaks to the customer and moves to the target location. But if a customer cannot explain his current location correctly or a proxy driver is not familiar with that position, a proxy driver cannot get to the customer quickly and the customer tends to be unsatisfied with the service. So the need for a system that provides proxy drivers with location information about customers and destination is rapidly increasing. Higher performance for embedded software with the intent of decreasing cost of hardware made widespread and development of hardware. Higher performance makes embedded software more complex, but to satisfy user demands and that of time-to-market, it has to be developed fast and achieve quality. in this paper, we shows the design and implementation of proxy driving service system using a location-based service using components. The experiment shows that the implemented system can provide efficient services to the customers and proxy drivers than existing systems.

Keywords: Component Based Development, Embedded Software Systems, Proxy Design, Location-based service.

Haeng-Kon Kim
Department of Computer information & Communication Engineering,
Catholic Univ. of Daegu, Korea
e-mail: hangkon@cu.ac.kr

Roger Y. Lee
Software Engineering & Information Technology Institute, Central Michigan
University, USA
e-mail: leelry@cmich.edu

Roger Lee (Ed.): SERA 2010, SCI 296, pp. 103–113, 2010.
springerlink.com   © Springer-Verlag Berlin Heidelberg 2010

1   Introduction

Complexity of embedded software is increasing due to the effect the customers demand in conjunction with the time constraints to get the product on market. Though the complexity of software is increasing and time to market is decreasing. With diminishing time allotted to creating embedded software the fact remains that most important factor in a product selling is to guarantee the quality of embedded software [1,2].

With wide spread of automobiles, the number of registered cars is rapidly increasing. To solve the traffic congestion, each country is today providing several forms of public transportation such as bus, the subway, and etc. But the majority of citizens who have cars of their own are still driving their cars. The use of private cars offers owner-drivers efficiency of movement, but the demerit is that the risk of traffic accident is high compared to other transportation means.

The proxy driving service is having a boom recently, which a proxy driver on behalf of a drunken driver has a car to the destination safely at night. To use a proxy driving service, a drunken driver only makes a call and gives his name, phone number, current location, and destination to a call center. On receipt of customers call, the call center sends a proxy driving request to all waiting proxy drivers. Proxy drivers have mobile terminals such as cellular phones or PDAs. If a proxy driver who is the nearest from the current position of a calling customer sends an accept message to the call center, then call center sends customers information to the designated driver.

But when the call center broadcast a request to all waiting proxy drivers, many proxy drivers may competitively response to a request. So it may be assigned to a farther proxy driver instead of nearer ones from the current location of the calling customer. In this case the waiting time of customer will grow longer and the problem is as follow: The designated driver tries to get to the customer much fast and the risk of traffic accident will become high.

After a proxy driver receives customer information and speaks to the customer to check the current location, he moves to the target position. But if a drunken customer cannot explain his current location correctly or a proxy driver is not familiar with that region, a proxy driver would call the customer or the call center several times to check the correct location. So the customer waits much more time than he has to, and will be unsatisfied with the service. To solve this problem, the need for a system that provides proxy drivers with location information about customers and destination is rapidly increasing.

Using a mobile phone based on Android platform supporting a location-based service and web application for a call center, this paper designs and implements a proxy driving service system which improvements the drawbacks of existing proxy driving service systems.

2   Related Works

2.1   *Embedded Software*

Embedded software is software developed for a specific purpose, this purpose is to satisfy the embedded systems needs[3]. What embedded software does in the embedded system are follows:

- Maximizes the embedded systems value.
- Offers interaction between outside in a high level.
- Performs interaction between human and computer.
- Supports harmonious operation between embedded systems in distributed environment.
- Provides features like security and reliability.

General software development is done using standard hardware and a common operating system (OS), but there are no absolute common hardware or OS in embedded software development. So developers have a greater flexibility in choosing these components. Which means the developer must understand hardware as well as software.

In early embedded systems, embedded software was a simple interface or program. Now embedded systems are developed in high efficiency programming languages and accomplish complex tasks.

2.2   *Embedded Software Testing*

Testing is a method to find defect(s) in software. Especially in embedded software development there are many hardware related limitations compared to generic software development[4]. Testing must work in a way that does not influence these limitations and must fit the following criteria:

- Real time testing: by definition, testing must be done in real time
- Non-interference of testing: testing must not interfere with elements not being tested.
- Support various kinds of connection methods: in order to patch or upgrade, an embedded system must support at least on means of support.

Thus, the embedded software developer must have knowledge about software and hardware to best construct and execute the appropriate test(s). Testing can be divided in a two forms: manual and that using automated tools. Using the manual method, hardware related knowledge are prerequisites and the testers knowledge and experience is also an important factor in testing.

Using an automated tool means using source code or a design model in predefined way to create a test case and tests automatically. Today complexity of embedded software makes manual based testing ineffective and requires lot of time so automated testing is taking place of it. Test cases are then sent to a target board using various methods and executed. Results of test case then are sent back to host and analyzed.

2.3 Embedded Software Quality and Evaluation

Use of embedded system is rapidly increasing. This increase of embedded systems creates many similar systems for the customer to choose from. This influences the quality required of embedded systems, because the system cant be improved or fixed when the development is over and fatal errors can affect the product, and consequently the evaluation of the company. Its because of consumer requirements that the importance of quality has been increased [6].

Developer/testers interest in quality is increasing to better serve the customer, because now quality is an influence on the sales of an embedded product. Quality evaluation results can view differently by the people who see it. So we need classification of elements.

Quality evaluation is another important factor which effects entire development process. Since evaluation collects element form entire development process, it can be used as resource for manual testing and reducing testing time.

2.4 Google Android

The Open Handset Alliance released the Google Android SDK on November 12, 2007[7]. The concept of Android platform is attracting more and more programmers in mobile computing fields. Android is a package of software for mobile devices, including an operating system, middleware and core applications. The Android SDK provides powerful tools and APIs necessary to develop applications on the Android platform using the Java Programming language.

Android platform is of open system architecture, with versatile development and debugging environment, but also supports a variety of scalable user experience, which has optimized graphics systems, rich media support and a very powerful browser. It enables reuse and replacement of components and an efficient database support and supports various wireless communication means. It uses a Dalvik virtual machine heavily optimized for mobile devices [8].

2.5 Adobe Flex

Adobe Flex is a software development kit released by Adobe Systems for the development and deployment of cross-platform rich Internet applications based on the Adobe Flash platform. Flex applications can be written using Adobe Flex Builder or by using the freely available Flex compiler from Adobe [9].

The initial release in March 2004 by Macromedia included a software development kit, an IDE, and a J2EE integration application known as Flex Data Services. Since Adobe acquired Macromedia in 2005, subsequent releases of Flex no longer require a license for Flex Data Services, which has become a separate product rebranded as LiveCycle Data Services.

3   Design Driving Service System

This section describes design issues of Proxy Driving Service System including the overall system architecture and the details of database design.

3.1   System Architecture

Figure 1 gives the overall system architecture of proxy driving service system. It is composed of three modules.

Call center (web application): Web application executes a remote method on the J2EE server through BlazeDS. Getting satellite coordinates for proxy drivers through Naver Map OpenAPIs from the server, the call center marks

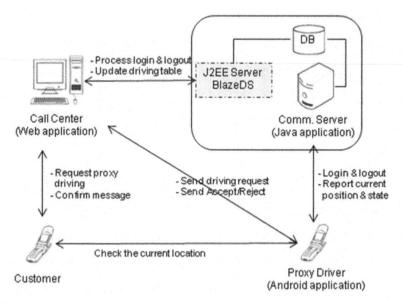

Fig. 1 The overall system architecture

their current location on the map and assigns a proxy driver who is the nearest from the calling customer among waiting drivers.

The translation between addresses and coordinates is performed by geocoding technique. Web application sends a driving request to a proxy driver, and updates a driving record table.

Android Application for a proxy driver: It is a socket client running on Android platform. Data exchange is performed between Android application and a call center through a communication server. Once a proxy driver logs in and reports its current location to a call center, he updates his location information every 500 meter movement from previous location using a LocationManager.

Communication Server: A communication server is the socket server implemented using Java language. It relays various data between Android application and a call center such as location information, login/logout requests, driving request/accept messages, etc.

3.2 Database Design

To implement a proxy driving service system, I designed a database that is composed of four tables. Figure 4 illustrates the detailed design of database.

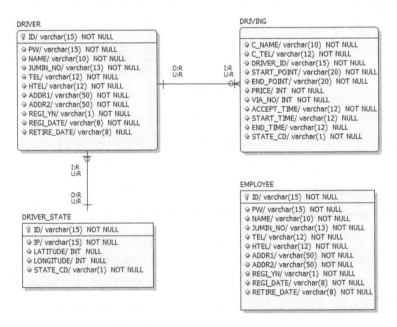

Fig. 2 ER Diagram of Database

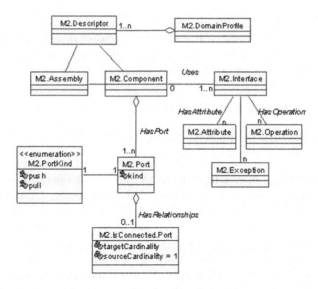

Fig. 3 Candidate Component of Location Services

(1) DRIVER table contains proxy drivers' private information.

(2) DRIVER-STATE table manages proxy drivers' state information. When drivers log in to a call center, the information about them is inserted and deleted on logout. A call center display waiting drivers on the map based on DRIVER-STATE table and sends driving requests to them.

(3) DRIVING tables contain proxy drivers' driving records.

(4) EMPLOYEE table manages employees' private information of a call center.

Figure 3 shows the candidate components to implement the systems. It consist of component, interface and some of attribute. It also has a provide and receive function to compose it later.

4 Execution of Location Service System

4.1 Implementation Environment

To implement the proxy driving service system using a location-based service, I used Apache Tomcat 6 as a web server program in a system installed Windows XP. I also used MySQL 5.0 as a database and implemented application programs using Java and Abode Flex.

Android provides Flexible map display, control functions and location support [7]. Android Location-Based APIs are used to collect user current position and display it on the screen, and Google Maps is used to display the current customer location on the mobile phone.

4.2 Implementation Details

This section describes the implementation details of the proxy driving service system.

Web application for a call center: A call center operator sees a login window through connecting a web site of the call center. Upon entering a user ID and password, web applications main window appears as shown in Figure 4.

On receipt of proxy driving requests of customers, an operator enters customers name and phone number, starting location, destination, connecting location, if any, and a fare. Then operator sends a confirm message to a customers mobile phone and assigns the request to a proxy driver who is the nearest from the current position of a calling customer among many waiting nearby drivers.

There are two methods to choose a proxy driver. That is, a direct selection or automatic selection. If any waiting driver accepts the request, then he sends an OK message to the call center.

After a driver selection, an operator clicks a request button, and then the request message is sent to a proxy drivers mobile phone.

Fig. 4 Web Applications Main Window

Fig. 5 The dialogue window to show a proxy driving request and The calling function in Android application

Android Application for mobile phones: When an operator sends a request message to the selected driver, a dialogue window to show the received request is appeared on the drivers Android application as shown in Figure 5.

If a driver chooses an OK button, then he will receive the detail information about the assigned request. A driver can talk over the mobile phone with the customer to check the current location of customer using the calling function in Android application as shown in Figure 5.

Meanwhile, the customers current position is marked as a starting location on the map in the drivers mobile phone based on Android platform using location-based service as shown in Figure 6.

Fig. 6 The dialogue window to show a proxy driving request and the calling function in Android application

As the driver gets to the customers current position, he starts a proxy driving for destination. On clicking a start button, the driver can see the destination marked on the map as shown in Figure 6.

As a driver arrives at the destination location, he clicks an end button on the map, and then the arrival information is sent to the call center. Finally, updated is the proxy driver state table and driving record table in the database based on the arrival information, and a proxy driving service is ended.

5   Conclusions

Using a mobile phone based on Android platform supporting a location-based service and web application for a call center, this paper designs and implements a proxy driving service system which improvements the drawbacks of existing proxy driving service systems. The experiment shows that the implemented system can provide efficient services to the customers and proxy drivers than existing systems.

Currently the implemented system chooses simply a proxy driver based on the distance between the customer and proxy drivers. But there is a possibility to apply several conditions to selection of a proxy driver. For example, customers may ask for a female driver, a male driver over forty years old, or a driver who is good at foreign languages, etc. In addition to, a call center provides a candidate driver list based on grades to customers, and customers have direct contacts with a proxy driver.

Further work is to apply above-mentioned functions to the implemented system and to improve it continuously.

Acknowledgements. This research was supported by the MKE(The Ministry of Knowledge Economy), Korea, under the ITRC(Information Technology Research Center) support program supervised by the NIPA(National IT Industry Promotion Agency)" (NIPA-2010-(C1090-1031-0001)).

References

1. Krstic, A., Lai, W.-C., Chen, L., et al.: Embedded Software-Based Self-Testing for SOC Design. In: Proceedings of the 39th conference on Design automation, pp. 355–360. ACM, New Orleans (2001)
2. Engblom, J., Girard, G., Werner, B.: Testing Embedded Software using Simulated Hardware. In: ERTS 2006, pp. 1–9 (2006)
3. El-Far, I.K., Whittaker, J.A.: Model-based Software Testing. Encyclopedia on Software Engineering, 1–22 (2006)
4. Sacha, K.: Evaluation of Software Quality. In: Software Engineering: Evolution and Emerging Technologies, pp. 381–388. IOS Press, Amsterdam (2005)
5. Kaner, C.: Architectures of Test Automation. In: Testing, Analysis & Review Conference (Star) West, San Jose (2000)

6. Deeter, K., Singh, K., Wilson, S., et al.: APHIDS A Mobile Agent-Based Programmable Hybrid Intrusion Detection System. In: Karmouch, A., Korba, L., Madeira, E.R.M. (eds.) MATA 2004. LNCS, vol. 3284, pp. 244–253. Springer, Heidelberg (2004)
7. Hessel, A., Larsen, K.G., Nielsen, B., et al.: Time-optimal Real-Time Test Case Generation using UPPAAL. In: Petrenko, A., Ulrich, A. (eds.) FATES 2003. LNCS, vol. 2931, pp. 118–135. Springer, Heidelberg (2003)
8. Artho, C., Barringer, H., Goldberg, A., et al.: Combining test case generation and runtime verification. Theoretical Computer Science, 209–234 (2005)
9. Open Handset Alliance, http://www.Openhandsetalliance.com/
10. Android - An Open Handset Alliance Project, http://code.google.com/intl/zh-CN/android
11. Wikipedia, http://en.wikipedia.org/wiki/Adobe_Flex

A New Compound Metric for Software Risk Assessment

Ahmad Hosseingholizadeh and Abdolreza Abhari

Abstract. Many different methods have been proposed for software risk analysis and assessment. These methods can be categorized in 3 groups: some methods are based on business owners and developers estimation about the probability and damage of a risk; some are based on software architecture analysis (using design diagrams), and some are based on source-code analysis. Each one of these approaches has some advantages and disadvantages, but none of them cover all risky aspects of a software project. The reason to this is that from one point of view software development is a heuristic process and it requires developers' heuristic analysis and opinions. But from another point of view there is a high probability that these opinions contain faulty evaluations. In this paper we propose an approach based on combining different metrics which are obtained from all three approaches of risk analysis. In our approach both Risk Probability and Risk Damage are obtained using this compound technique. The architectural risk of the components is calculated based on the cyclomatic complexity of the statecharts; the source-based risk is obtained by a code weight association technique and these values are aggregated with the analysts opinions to produce the risk model. We provide a case study to present the results of our approach.

1 Introduction

According to [7] risk is defined as the function of the possible frequency of occurrence of an undesired event, the potential severity of resulting conse-quences, and the uncertainties associated with this frequency and severity.

Ahmad Hosseingholizadeh
Department of Computer Science, Ryerson University, 350 Victoria Street,
Toronto, Ontario M5B 2K3, Canada
e-mail: ahossein@ryerson.ca

Abdolreza Abhari
Department of Computer Science, Ryerson University, 350 Victoria Street,
Toronto, Ontario M5B 2K3, Canada
e-mail: aabhari@ryerson.ca

Roger Lee (Ed.): SERA 2010, SCI 296, pp. 115–131, 2010.
springerlink.com   © Springer-Verlag Berlin Heidelberg 2010

Risk in the context of software engineering is defined as the probability that a software development project experiences unexpected and inadmissible events such as termination, wrong budget and time estimations, poor quality of the software solution, wrong used technology, etc [1]. Experience indicates that 80% of all the potential for a software project failure can be account for by only 20% of the identified risks [3]. Choosing an optimal strategy to rank risks and identify this 20% will have a deep effect on expenses reduction and product's functionality and it allows the organization to take early mitigating actions to detect the defects of high risk components [12].

A subset of the software risks is related to the poor quality of the solution and probability of the operation failure of its components. Finding the problematic components which are more probable to fail can guide the project team to plan an optimal development which minimizes this probability and its effects[6]. Also risky components should be identified to be developed and tested in the early stages of the development to minimize risks of the project [2]. In this paper we use the word *Risk* to refer to this special subset of the software risks. Software risk can be quantified as a combination of the following two metrics [14]:

- Risk Probability
- Risk Damage

Different approaches have been proposed to measure these 2 metrics. Some of them are based on *business owners and developers opinions* about the developing software product. In these approaches components' riskiness[1] is estimated by developers, and component failure's damage is estimated by developers and business owners collaboration. Some of the important factors that affect the result of this approach are: developers and business owners' experience, precise insight of the problem domain, etc.

Some other approaches are based on *formal and computable techniques* which are possible to be programmed and automated. This means that risk analysis can be done by means of an analyser application. These techniques can be categorized in two groups: The first group are those techniques that can process *the architectural design and modeling artifacts* of a component (e.g. *Statecharts* or *Sequence Diagrams*) and calculate different metrics such as *Static* or *Dynamic Complexity* to generate the *Risk Model* of each component and estimate its risk and reliability [4][5][13][14][15]. The second group are those techniques that process *the source-code of a component* and estimate the risk metrics based on the code specifications such as the number of conditional statements, function calls, etc [8][9]. These values will be used to determine the risk factor of components and produce the the Risk Model of the software product. The Risk Model can be used to manage the debugging and testing strategies of a development process.

---

[1] We use the term *riskiness* to refer to relative probability of failure of a component in relation to other components of the same software product. This value doesn't identify the absolute probability of failure.

Each one of these three techniques have some advantages and disadvantages. The *owners and analysts opinion* is a very important factor to determine the risky components, because risk discovery is a heuristic process and no two projects are exactly alike. But this type of analysis cannot be enough because many of the risks are not determinable by analysts due to lack of experience, unexplored areas of new technologies that are used in every different project, implementation details which are not clear before the design and coding phases begin, weak points of development team in the area of the new technologies, etc. In addition to these, there is always the human fault factor which causes some risks to remain undetermined.

Architectural analysis techniques are not enough either because an application with a simple architecture can have a very complicated internal logic in the body of the functions and procedures. These risks can only be determined by code analysis techniques.

Also just using *Code-Based analysis techniques* will not generate a precise Risk Model because the internal structure of functions and procedures of an application can be simple, but the relation between its components can be very complicated.

Considering the above and different phases of a development process (design and implementation of each partial system) which take place before the test, it can be seen that a risk analysis approach can produce a reliable model only when it uses the information obtained from each of the analysis, design (architecture) and implementation (code) phases. In our proposed approach owners and analysers' estimation is considered as one of the effective factors and it is combined with the values that are obtained from Architectural and Source-Code analysis which results in a compound model that takes all three aspects of a risk analysis into consideration.

$$RiskMetric = f(AnalystsEstimation, ArchitecturalRisk, SourceBasedRisk) \qquad (1)$$

The rest of the paper is organized as follows. First we introduce our proposed technique to calculate the riskiness (probability factor) of software components based on the mentioned three viewpoints. In section 3 we will present our proposed technique to predict the risk damage which is also based on the mentioned viewpoints. In section 4, we present an approach to combine the calculated values and create a risk table called Compound Risk Table which is used to categorize software components based on their risk factor. In section 5 we present a case study to show the results of applying this approach to a software project. Finally in the last section the future research plan to achieve a more applicable and precise approach will be presented.

2 Riskiness of Components

In this section we introduce our proposed approach to determine the riskiness (relative probability of failure) of a component of an application. Our goal is to define a technique to compare the risk of different components of a software product. As discussed in the previous section our approach is based on a compound metric which takes all aspects of a project into consideration. First we introduce the elements of our compound metric and their calculation method, and after that we present our approach to aggregate these elements and calculate the compound metric.

2.1 The Elements of the Riskiness Compound Metric

To determine the relative probability of failure of each component (in relation to other components of a software product) we will calculate its riskiness factor by considering the following viewpoints:

- Project Analysts and Managers' estimation about the probability of failure of each component
- Architectural Analysis of each component
- Source-Code analysis of each component

2.2 Component Riskiness Based on Analysts Estimation

Project managers and analysts can use their experience to determine the relative riskiness of each component. In order to do this, after each design session analysts and project managers will discuss each component of the designed partial-system and associate a number between 0 to 100 to each component which identifies its riskiness (The higher the number is, the more probable is the failure of the component). We will show this value by CR_1.

In short term: $CR_1(k) = The\ riskiness\ factor\ of\ the\ component\ k\ which\ is\ determined\ by\ the\ project\ analysts\ and\ managers.$

2.3 Component Riskiness Based on Architectural Analysis

In 1976 McCabe proposed a technique to determine the complexity of an application[10]. In his method a new metric called *Cyclomatic Complexity* was introduced which determined the complexity of an element of an application by evaluating $V(G) = E - N + 2$ using the element's *Control Flow Diagram*. In this equation $V(G)$, E and N stand for *Cyclomatic Complexity*, *number of Edges* and *number of Nodes* respectively. In [4] this definition of complexity has been extended to the software architecture level. In their proposed approach a *Statechart* is designed for each component which is used instead of the *Control Flow Diagram* to make it possible to calculate the complexity in the Architectural level. This complexity has been used as the

Architectural Risk Factor of a component. We use the same approach to calculate the architectural complexity.

In order to determine the Architectural Risk, the statechart of each component is designed based on each use case; then cyclomatic complexity is calculated based on this diagram (using the number of edges and nodes). In our study we look at each component from a general point of view, thus we don't design separate statecharts for different use-cases; instead we propose to design only one statechart which contains one default or idle state and all the other possible states of a component based on all use-cases. The Complexity Factor of component k will be calculated using the following equation:

$$CR_2(k) = E(k) - N(k) + 2 \qquad (2)$$

In this equation $E(k)$ and $N(k)$ represent the number of edges and nodes for the statechart of component k, and $CR_2(k)$ stands for the Cyclomatic Complexity of component k which represents its riskiness and subsequently its relative failure probability factor. In section 2.5 this value will be combined with other metrics and normalized to produce a compound metric that can be used to determine the riskiness of a component.

2.4 Component Riskiness Based on Source-Code Analysis

In this section we describe our proposed approach to determine the riskiness of a component based on source-code analysis. There are different approaches which use different factors to calculate the source-code's riskiness: number of lines of code, number of data accesses, cyclomatic complexity (based on control flow), etc. In this section we propose a new approach which uses the basic idea of [8] and improves it to have a better estimation of the riskiness of the analysing component. In [8] *Static Risk Model* based on *summation* scheme is defined as following:

$$V * \alpha + F * \beta + D * \gamma + C * \epsilon + P * \rho \qquad (3)$$

In this formula, V, F, D, C and P are metrics which are extracted from the code. V stands for number of variable definitions, F number of function calls, D number of decisions, C number of c-uses[2] and P number of p-uses[3]. $\alpha, \beta, \gamma, \epsilon,$ and ρ are the weighting factors which are used to give either more or less emphasis to the metric components of the above formula.

---

[2] c-use(computational use) is calculated for each block of code as follows: the number of variable usages in the right hand side of each assignment statement, plus the number of variable usages in output commands.

[3] p-use(predicate use) is calculated for each block of code by counting the total number of variable usages in conditional statements.

Our proposed technique is an improved version of the former logic. We describe our approach using the following two sample procedures shown as *Procedure 1* and *Procedure 2*.

Procedure 1 - A procedure with one statement in the *if* body

```
Procedure foo()
{
    if (condition 1)
    {
        statement 1;
    }
    statement 2;
    statement 3;
    statement 4;
}
```

Procedure 2 - A procedure with 3 statements in the *if* body

```
Procedure bar()
{
    if (condition 1)
    {
        statement 1;
        statement 2;
        statement 3;
    }
    statement 4;
}
```

Using the formula (3) these two procedures will have the same value for risk; however by further analysis it can be seen that *Procedure 2* is more risky than *Procedure 1*. The reason for this is that if in *procedure 2* as an example the *condition 1* fails to operate properly, more statements will be executed which are not supposed to be executed; thus it can be said that *condition 1* in *Procedure 2* is more damaging than *condition 1* in *Procedure 1*. This concept can be interpreted by saying that the riskiness of *condition 1* is added to the riskiness of *statements 1, 2* and *3* in *Procedure 2* and made them more risky (As an example, it can be said that the riskiness of *statement 2* in *Procedure 2* is based on two factors, first the risk of failure of *statement 2* itself and second the risk of failure of proper execution of *condition 1*). With this approach the effect of the risk of *condition 1* can be interpreted as an increase in the riskiness of its corresponding block of statements.

The former point is valid for all the statements in conditional/loop blocks (such as *for*, *while*, etc.). Having a bug in the condition of a loop will effect its whole body by the measure of the number of its wrong executions; thus the risk factor affects the body of the loop in a more severe matter. It is more complicated to measure this effect in loop statements because in some cases the number of executions of their body is not determinable prior to the application's execution. So to make it simpler we consider a constant weight to be assigned to all the loops which should be determined by the developers and project analysts.

Considering the former points, we change the formula (3) into the following equation:

$$BR(n) = V' * \alpha + F' * \beta + D' * \gamma + C' * \epsilon + P' * \rho + \sum BR \qquad (4)$$

In this equation $BR(n)$[4] stands for riskiness of block n, and $\sum BR$ stands for the sum of all BRs of the internal blocks of block n. If we represent all the metrics V, F, D, C and P with X, all X's in (5) are defined as follows:

$$X' = X * BRF(n) \qquad (5)$$

In this formula $BRF(n)$ represents the *Block Risk Factor* of the block n. We define $BRF(n)$ as following: In the first level of each procedure (or function) BRF is 1; by entering each if statement block, BRF would be equal to the BRF of if statement's parent block incremented by 1. BRF of each loop block is equal to BRF of loop's containing block plus 5 (this is our proposed value and our approach can be applied with other values for different projects based on their implementation specifications). To explain this technique we provide the example shown in *Procedure 3*.

Procedure 3 - A sample procedure

```
Procedure example()
{
    int a, b;
    int max;
    cin >> a >> b;
    if (a > b)
    {
        max = a;
        cout << max;
        Proc2();
        for (int i = 1; i <= max; i++)
            Proc3();
    }
    else
    {
        cout << b;
    }
    cout << a * b;
}
```

In *Procedure 3* for the outer most procedure, BRF is 1. For *if* and *else* blocks BRF is equal to 2 (BRF of the1 parent block + 1) and for the level 3 block which is inside the *for loop* BRF is equal to 7 (BRF of the containing block + 5). In the inner most block (*for loop*) there are 2 p-uses ($i <= max$),

---

[4] Block Risk.

a c-use $(i++)$ and a function call $(Proc3())$. By assuming that all the weights are 1:

$$BR(4) = 1 * 7 + 2 * 7 + 1 * 7 + 1 * 7 + 0$$

in the *if* block there is a variable definition (since *int i* is executed only once we consider it outside the *for loop* body), 2 c-uses (*cout << max* and *max = a*), a function call $(Proc2())$ and a nested block $(BR(4))$:

$$BR(3) = 1 * 2 + 2 * 2 + 1 * 2 + BR(4)$$

With the same approach, BR(2) for the *else* block and BR(1) for the procedure's block are calculated as follows:

$$BR(2) = 1 * 2 + 0$$
$$BR(1) = 3 * 1 + 2 * 1 + 2 * 1 + BR(2) + BR(3)$$

Finally the riskiness of the outer most procedure $(BR(1))$ is equal to 45. The riskiness of a component is the sum of the riskiness of all the procedures inside it:

$$CR_3(k) = \sum BR(i) \tag{6}$$

This metric is a relative metric which can only be interpreted in relation to other components riskiness. We also defined another metric called CAR which stands for Component Average Risk:

$$CAR_3(k) = \frac{\sum BR(i)}{N} \tag{7}$$

CAR_3 stands for the average of riskiness of all the methods inside the component k; N is the number of methods inside a component (The index 3 in CR_3 and CAR_3 is used to distinguish these metrics from other metrics which were introduced in the previous sections). If a component has a relatively bigger CR_3 then it can be said that this component is riskier than others; having a component with a relatively bigger CAR_3 means that the density of risk in this component is higher.

2.5 Combining the Metrics

In this section the purpose is to combine the acquired metrics from the previous sections to generate the new compound metric. We will combine CR_1, CR_2 and CR_3 and create the new metric CR which will determine the Component's Risk. By comparing CRs of different components in a project, one can identify the most risky components. In the first step we normalize the values of CR_is using the following formula:

$$NCR_i(k) = \frac{CR_i(k)}{\sum_k CR_i(k)} \tag{8}$$

In (8) $NCR_i(k)$ stands for *the normalized riskiness of component k*. The parameter i identifies each one of the 3 metrics which where calculated in the last three sections. $\sum_k CR_i(k)$ represents the sum of CR_is of all the components. Using (8) we will normalize all the calculated values for all three metrics and the result will be three values of $NCR_1(k) \sim NCR_3(k)$ for each component of the developing application. NCR_1 stands for normalized riskiness estimated by *project managers and analysts*, NCR_2 stands for normalized riskiness calculated by *Architectural Analysis* and NCR_3 stands for normalized riskiness calculated by *Source-Based Analysis*. After normalization, the compound metric CR (Component Risk) is calculated using the following formula:

$$CR(k) = \theta * NCR_1(k) + \omega * NCR_2(k) + \sigma * NCR_3(k) \qquad (9)$$

In this formula $CR(k)$ stands for the riskiness of component k, $NCR_i(k)$ stands for the normalized riskiness of the three mentioned approaches and θ, ω and σ are the weighting factors which are used to give more or less emphasis to each of the metric elements. Considering that this formula uses all the aspect of a software project, we believe that the classification of components based on CR(k) is mush less faulty than other methods. This method can be used in any stage of a software development. If the risk analysis is performed at the early stages of development (before any implementation) CR(k) can be calculated by putting 0 as the value of NCR_3; in this way CR(k) can be calculated without any implementation and since the value of CR for one component is a relative value and it is eventually used by being compared to CR of other components, the obtained values can be used without any change in the original algorithm.

3 Risk Damage Analysis

In this section we will describe our proposed approach to determine the damages that each risk can cause in a software product. Using our approach, components can be classified based on their potential damage. We will determine the components' failure damage based on the project's *Analysts and Managers estimation* of the damage of each component and *Architectural Analysis* of the software structure.

3.1 Component Failure Damage Based on Analysts Estimation

The logic of the application is a very important factor in a component failure damage. Analysts take the logic of the applications into account while they are estimating the potential failure damage of a component; thus it can be seen that analysts and managers estimation of the wrong execution of a component is very important. The most important resources that analysts can use to clarify the business logic of the developing application are the business owners; but the problem is that usually owners don't have enough understanding of the internal structure of a software and a middle step should be taken to map owners' view of the system to the actual structure of the application. In the context of Risk Analysis, the operations which are important to the business and their responsible components should be identified. Our proposed method for this mapping is as follows:

In the early sessions of project analysis which take place between business owners, analysis and developers, use-cases of the developing application should be discovered and identified. After this identification business owners should assign a value between 0 to 100 to each use-case which determines the importance of that use-case. A use-case with the value of 100 would be the most important use-case. Considering this value, analysts and developers can discover the participating components in each use-case and identify the most important components of the software. Knowing that it is not possible to involve business owners in the structural design of the application, this approach helps the developers and analysts to map owners' view to their design. In this approach the importance of each component will be shown by **imp(k)** which stands for the importance of component k extracted by analysts and developers from importance of use-cases. Since $imp(k)$ is determined in the first steps of the development, as the application is being developed its architecture will be modified, thus developers might need to re-assign or modify the values of $imp(k)$ for different components.

We define the *Internal Operation Failure Damage* of component k as follows:

$$IOFD(k) = \frac{imp(k)}{100} \tag{10}$$

In this equation $IOFD$ would have a value between 0 and 1; higher values represent more damaging components.

3.2 Architectural Damage Evaluation and Damage Metric Aggregation

In this section we introduce a new factor regarding the failure damage of software components. In our proposed method the failure damage depends on the *Dependencies* between components. To clarify this first we introduce the *Depth* concept. Having the package diagram of an application, the *Depth of a package* is defined by assigning numbers to the packages of the application starting from 1 which is assigned to the upper (outer) most package and moving down assigning 2, 3, ... to the packages in the lower levels, and the *Depth of a component* is defined as the number assigned to its containing package (Fig.1).

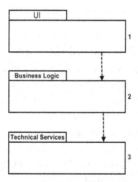

Fig. 1 Depth of each component is determined by the depth its container package

Since in a package diagram the functions of the higher components depend on the correct execution of the lower components, therefore in case of a failure if a component has a higher depth value it will cause more damage. As an example if there are multiple components in the second tier of the package diagram (depth : 2) which are all dependent on one component in the 3rd tier, a failure in the operation of the lower component can potentially cause the failure of operation of all the higher components. This dependency can be between components of the same package as well. Thus we can say that those components which have more dependent elements cause more damage. Considering this dependency factor, we define the failure damage of a component as following:

$$CFD(k) = IOFD(k) + \tau * DCFD(k); \tag{11}$$

In this equation $CFD(k)$ stands for *Component Fail Damage*, $IOFD(k)$ stands for *Internal Operation Fail Damage* and $DCFD(k)$ stands for *Dependent Components Fail Damage* of component k. τ is the weighting factor. Using equation (11) the effect of the architectural design of the components of an application is also considered in the failure-damage estimation. $DCFD$ of a component is obtained by calculating the sum of CFDs of all its dependent components. This process should start from the upper-most tier (depth 1). $DCFD$ for the components in the upper-most tier which have no dependent elements is 0, thus in this case all CFDs are equal to $IOFD$s which are extracted from analysers and owners estimation of the failure damage and have a value between 1 and 0. In this step 0 means *not damaging* and 1 means *very damaging*. After the first tier CFDs of the second tier can be calculated by adding $IOFD$ of each component and sum of CFDs of its dependent components $(DCFD)$.

By comparing the value of CFD for different components the most damaging components can be identified.

4 Compound Risk Table

In this section we will use the values of CR and CFD to create a table called the *Compound Risk Table*. The rows of this table can be sorted based on the Failure Damage and Riskiness (probability of failure) to determine the most harmful and risky components. This table can be used to identify the most damaging and risky components and as a guide for development scheduling and software test and debug management.

The value of CR is obtained by adding up three NCR_is and since each NCR_i has a value between 0 and 1, therefore the value of CR is a number between 0 and 3. According to this, the risky components are those with a CR closer to 3 than other components.

Since the number of tiers of a software product is different among different applications, the boundaries of CFD can not be pre-identified. If different components have so many dependent elements in the higher tiers the value of $DCFD$ will increase very fast and as a result the value of CFD can raise up to big numbers. The value of CFD will be used to classify the components of an application in four categories:

- Negligible
- Marginal
- Critical
- Catastrophic

To find the membership of different components in these categories the highest value of CFD will be identified and divided into 4 equal number ranges corresponding to the four categories of damage (the range with lowest CFDs identify the category *Negligible* and higher CFDs identify the *Marginal, Critical* and *Catastrophic* categories respectively). The class of the components can be determined by identifying the membership of each of the components to these groups. Having CR, CFD and *Damage Class* of all the components, the *Compound risk table* will be created as shown in table 1.

Table 1 Example of a Compound Risk Table

	CR	CFD	Damage Class
Component 1	CR(1)	CFD(1)	[DamageClass]
Component 2	CR(2)	CFD(2)	[DamageClass]
...	...	...	...
Component N	CR(N)	CFD(N)	[DamageClass]

5 Case Study and Results

In this section we present the results of applying our approach to a sample project which is a regression analysis system based on Genetic

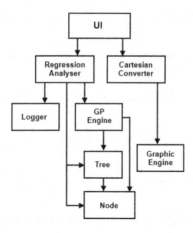

Fig. 2 Dependency Diagram

Programming. Figure (2) shows the components of this sample project and their dependencies.

The obtained values for $CR_1, CR_2, CR_3, IOFD$ and CFD are shown in table 2. In this example all the weighting factors (except τ which is associated with the value 0.2) are given the same value of 1. The reason for choosing 0.2 for τ is that the dependency between components are not total, meaning that only a fraction of the operations of the component under analysis are dependent on other components. In our study we chose $\tau = 0.2$ which is based on our estimation of the average role of dependencies in the operation of the components.

Considering the values in table 2 the Compound Risk Table for this example is shown in table 3 and the compound risk diagram which is generated based on the compound risk table is shown in figure (3).

By looking at table 3 it can be seen that the component which is more probable to fail is the *GP Engine*. Considering that from the damage point

Table 2 Metric Components

	CR_1	CR_2	CR_3	$IOFD$	CFD
UI	50	3	373	0.5	0.5
Regression Analyser	60	8	362	1	1.1
GP Engine	80	10	1068	1	1.22
Tree	20	4	64	1	1.46
Node	10	4	28	1	1.76
Cartesian Converter	60	10	231	0.4	0.5
Graphic Engine	40	6	45	0.5	0.6
Logger	20	2	41	0.1	0.32

Table 3 Compound Risk Table

	CR	CFD	Damage Class
Logger	0.12	0.32	Negligible
UI	0.38	0.5	Marginal
Cartesian Converter	0.49	0.5	Marginal
Graphic Engine	0.27	0.6	Marginal
Regression Analyser	0.51	1.1	Critical
GP Engine	0.93	1.22	Critical
Tree	0.17	1.46	Catastrophic
Node	0.12	1.76	Catastrophic

of view *GP Engine* is classified as *Critical*, a good test plan should start by
testing this component. Components *Tree* and *Node* are from the class *Catastrophic* but they have a very low relative probability of failure, thus in the test
plan they are not considered as the highest priority. By looking at this table
project manager can have a precise understanding of the risk specifications
of different components and plan an optimal test and development.

To verify the results of our approach, after the initial implementation we
put the application under test and measured the number of bugs of each
component. The result of this initial test can be seen in table 4.

As it can be seen in table 4 the highest number of bugs were found in
the *GP Engine* which is also the most risky component (highest *CR*). The
result of the test also matches with the obtained value for the second most
risky component which is the *Regression Analyser*. The number of bugs for
Cartesian Converter is slightly different from the result of the risk analysis.
The reason for that is because of the fact that this component is from a
type which is more commonly developed and known by the developers of the
application. In large projects the components of the subsystem under analysis

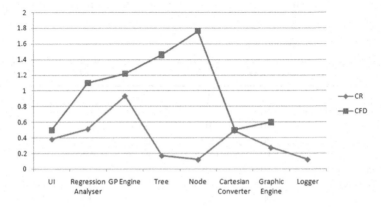

Fig. 3 Compound Risk Diagram

Table 4 Comparison of CR and number of bugs

	CR	Number of Bugs
Logger	0.12	1
UI	0.38	1
Cartesian Converter	0.49	1
Graphic Engine	0.27	2
Regression Analyser	0.51	5
GP Engine	0.93	6
Tree	0.17	0
Node	0.12	0

would most likely be of the same type, thus the effect of this issue would be much less and almost negligible.

Our proposed approach can be used at any stage of the development of the software. In the early stages risk analysis can be done by excluding NCR_3 from the calculations; thus the risk metric will be calculated without considering code-dependent parameters. This risk metric can be used for basic planning and task association in the early stages of the development. After each subsystem development in each development cycle, code-dependent metrics (NCR_3) can be included in the Risk Analysis and as a result by identifying risky components, subsystem test and debug can be performed more precisely and more effectively. The high flexibility of our proposed approach (which is a result of using the weighting factors in calculating the metrics) makes it adjustable for different development teams with different experiences and skills.

Most of the proposed metrics in our approach are analytical metrics, meaning that there is a specific calculative approach to obtain them. This makes it possible to create an analysing tool which can be used to analyse a software and generate the metrics. This analyser tool can be designed to get the developed code and architectural diagrams as inputs and produce their corresponding risk factors (NCR_2 and NCR_3) and combine these values with the analysers' estimated metrics (NCR_1) to generate the Risk Model of the developing software. This will cause the risk analysis task to be very applicable and easy which results in a practical and precise approach to assist the creation of the development plan.

6 Conclusion and Future Research

In this paper we discussed that software project risks can be analysed from different points of view. We argued that each of these approaches identify the risk of a component by considering a subset of the factors that affect the quality and reliability of each component, and in order to have a more precise and reliable risk model we have to take all these factors into account which resulted in our compound risk model. We showed that our new technique can assist the project managers to plan the development by considering the risk factor of different components of the application which can be obtained by analysing the software project from different aspects.

For our future research we want to complete our code-analysis approach by including the logics of the application into our calculations. We also want use intelligent techniques in our code analyser and apply a training strategy which makes it possible to identify those kinds of risks associated with the development characteristics of each specific development team. Additionally we want to design a technique to be added to our tool to assist debugging of the identified risky components using the obtained risk information and intelligent automatic debugging techniques (Genetic Programming Automatic Debugging techniques, etc). Our long term goal is to design an approach to assist identifying the design problems of a given software and propose an alternative better design. To do that, our plan is to study software design patterns and examine their potentials to provide risk information about each component.

References

1. Tao, Y.: A Study of Software Development Project Risk Management. In: Proceedings of the 2008 International Seminar on Future Information Technology and Management Engineering, pp. 309–312 (2008)
2. Larman, C.: Applying UML and Patterns: An Introduction to Object-Oriented Analysis and Design and Iterative Development, 3rd edn. Addison Wesley, Reading (2004)
3. Pressman, R.S.: Software Engineering: A Practitioner's Approach, 5th edn. McGraw-Hill, New York (2001)
4. Popstojanova, K.G.: Architectural-Level Risk Analysis Using UML. IEEE Transactions on Software Engineering 29(6), 946–960 (2003)
5. Yacoub, S.M., Ammar, H.H.: A Methodology for Architecture-Level Reliability Risk Analysis. IEEE Transactions on Software Engineering 28(6), 529–547 (2002)
6. Khoshgoftaar, T.M., Seliya, N., Liu, Y.: Genetic Programming-Based Decision Trees for Software Quality Classification. In: Proceedings of 15th IEEE International Conference on Tools with Artificial Intelligence, pp. 374–383 (2003)
7. NASA Safety Manual NPG 8719.13A, Software Safety (1997)
8. Wong, W.E., Qi, Y., Cooper, K.: Source Code-Based Software Risk Assessing. In: Proceedings of the 2005 ACM symposium on Applied computing, pp. 1485–1490 (2005)
9. Deursen, A., Kuipers, T.: Source-based software risk assessment. In: Proceedings of the International Conference on Software Maintenance, p. 385 (2003)
10. McCabe, T.J.: A Complexity Metrics. IEEE Transactions on Software Engineering 2(4), 308–320 (1976)

11. Khan, S.: An approach to facilitate software risk identification. In: Proceedings of 2nd International Conference on Computer, Control and Communication, pp. 1–5 (2009)
12. Emam, K.E., Melo, W.: The Prediction of Faulty Classes Using Object-Oriented Design Metrics. Technical Report NRC 43609, Nat'l. Research Council Canada, Inst. For Information Technology (1999)
13. Bass, L., Nord, R., Wood, W., Zubrow, D.: Risk Themes Discovered through Architecture Evaluations. In: Proceedings of the Working IEEE/IFIP Conference on Software Architecture, pp. 1–10 (2007)
14. Cortellessa, V., Popstojanova, K.G., Appukkutty, K., Guedem, A.R., Hassan, A., Elnaggar, R., Abdelmoez, W., Ammar, H.H.: Model-based performance risk analysis. IEEE Transactions on Software Engineering 31(1), 3–20 (2005)
15. Cheung, L., Roshandel, R., Medvidovic, N., Golubchik, L.: Early prediction of software component reliability. In: Proceedings of the 30th international conference on Software engineering, pp. 111–120 (2008)

Towards a Tool Support for Specifying Complex Software Systems by Categorical Modeling Language

Noorulain Khurshid, Olga Ormandjieva, and Stan Klasa

Abstract. Formal methods are proven approaches to ensure the correct operation of complex interacting systems. However, current formal methods do not address well problems of verifying emergent behavior and evolution, which are two of the most important characteristics of complex software systems. A subset of the Category Theory has been proposed in this paper to specify such systems. The categorical modeling language depicted here is the first step towards a powerful modeling paradigm capable of modeling emerging and evolving behavior of complex software. The approach is illustrated with a case study of Prospecting Asteroid Mission (PAM) from the NASA.

Keywords: Formal methods, categorical modeling language, complexity, Prospecting Asteroid Mission (PAM).

1 Introduction

The ever increasing software systems' complexity curve is one of the few challenges the software industry has been confronting to date. Representation of complex structures and behavior, either diagrammatical (such as UML) or mathematical (for instance, VDM, Z, etc.) is in turn complex too. Breaking a big problem into smaller problems is the orthodox human mind proposition. But how to keep track of smaller problems, as to where they fit into the bigger problem is a question.

Category theory suggests a solution where abstraction is employed as required while focusing on a specific sub-problem coming to the level of that problem, the bigger problem still in view. Software engineering is very fragmented, with many different sub-disciplines having many different schools within them. Hence, the kind of conceptual unification that category theory can provide is badly needed. As pointed out by Goguen [5], category theory can provide help in dealing with

Noorulain Khurshid, Olga Ormandjieva, and Stan Klasa
Computer Science & Software Engineering Department, Concordia University
1515 St. Catherine St. West, Montreal, Quebec, Canada H3G 1M8
e-mail: {n_khursh,ormandj,klasa}@encs.concordia.ca

Roger Lee (Ed.): SERA 2010, SCI 296, pp. 133–149, 2010.
springerlink.com

abstraction and representation independence. In engineering complex systems, more abstract viewpoints are often more useful, because of the need to achieve independence from the overwhelmingly complex details of how things are represented or implemented.

The proposed in the paper Category Modeling Language (CML) is based on category theory constructions. It provides a mathematically rigorous yet natural modeling of complex software systems as if there were a single structure guiding its behavior, where the whole system is modelled as autonomic colonies of interacting objects in their certain neighbourhoods. The systems exhibit autonomic emergent behavior and evolving behavior, commonly referred to as "self-*" properties. CML can be used to present the categorical modeling of self-* properties for specifying autonomic structure and behavior, and a graphical tool to support such modeling. NASA's Prospecting Asteroid Mission (PAM) [11] has been used as a case study in this paper to demonstrate the abilities of the CML.

The rest of the paper is organized as follows: section 2 presents the research problem and our approach. Section 3 reviews briefly the related work. Section 4 provides a very brief grounding in the category theory constructs used in the rest of the paper. The case study (PAM) is described in section 5 and section 6 includes the CML and its illustration on the PAM. The tool support is presented in section 7. The conclusions and the future work directions are outlined in section 8.

2 Problem Statement

Our background study into existing modeling methodologies applicable to complex software systems has shown that the diversity of elements (structural, behavioral, etc.) involved in software engineering can be only dealt within a framework of a powerful modeling paradigm. Our approach is to develop complex software systems' modelling language based on category theory that should possess the following characteristics:

- *Semantical power.* The formalism should provide rich semantics close to the real world semantics. The semantics have to be formalizable to rule out ambiguities, and to allow for computer assisted processing.
- *Expressive power.* A wide range of semantic constraints (system policies) usually changeable and subjected to unpredictable evolution should be expressible.
- *Abstraction flexibility.* Depending on the particular usage the framework should allow to abstract away from unnecessary detail. On the other hand, modeling of different types of artifacts should be done in a uniform way.
- *Unifying (Simulation) ability.* In order to support currently available and used modeling formalisms we require our modeling framework to have the possibility to simulate a wide range of other modeling formalisms.
- *Comprehensibility.* Since primary creators and users of complex system models are people the formalism should be easily comprehendible.

There are two main components of comprehensibility, namely, graphical syntax and modularity. Thus, Category Theory is our choice of the visual formalism, which offers techniques for manipulating and reasoning on diagrams for building the hierarchies of system complexity, and allow systems to be used as the components of more complex systems, hence making it possible to infer the properties of the systems from the sub-systems' configurations.

3 Related Work

The software modeling approaches currently used fail to integrate these characteristics into a single formal framework in the scope of complex systems. For instance, statecharts has been one of well-known modeling approaches for specifying complex systems. The author in [18] presents an extension of the conventional statecharts to visualize complex systems. The idea being, small diagrams can express complex behavior. In [19] a higraph-based language of statecharts has been used for knowledge representation and behavioral specification of complex systems. Nonetheless, in a complex system the exponentially growing multitude of states can become unmanageable before long. UML has been one of the widely used modeling methodologies for modeling software architecture to software design. In paper [20] the author presents UML's extensibility mechanism by defining UML-based constructs to model real-times systems' software architecture. Although capable of modeling a software system at different levels of generalization, UML lacks the abstraction level which could be achieved by the use of Category Theory to model heterogeneous characteristics of the system such as functional and non-functional properties using the same constructs.

Despite the popularity of category theory in some fields of computing science, very few applications in the field of software engineering can be found in the literature. The only published work on modeling autonomous systems using category theory [8] served as the structure for the research presented in this paper. Its author stated that an autonomous system is a group of cooperating subsystems, and defined a specification language for such systems based on category theory. A constructor for communication by using monoids was introduced, and the feasibility of categorical approach was proven, but no systematic methodology was proposed. In [8], category theory for software specification adopts the correct by construction approach, in which components can be specified, proved, and composed so as to preserve their structures. The authors in [7, 14] advocate the use of the category theory for axiomatizing, differentiating and comparing structures in Multi-Agent Systems (MAS) where categorical representation takes its formal meaning and carries with it the intuition that comes from practice.

There is also some related work regarding our case study. The paper [9] states a formal task-scheduling approach and model the self-scheduling behavior of PAM by an autonomic system specification language. The authors in [15] summarize necessary properties for the effective specification and emergent behavior predication of PAM. They also compared current formal methods and integrated formal methods for the specification of intelligent swarm systems with the emergent behavior. However, there is no single formal method satisfying all the required properties for specifying PAM, and the PAM specification cannot be easily converted to program code or used as the input for model checkers when using integrated formal methods.

4 Category Theory

Category theory is an advanced mathematical formalism that is independent of any modeling or programming paradigm, which is adequately capable of addressing "structure" [2,3,4]. In mathematics, category theory is an abstract way of agreement between various mathematical structures and relationships between them. Everything is abstracted from its original meaning to categorical meaning. For example, sets abstract to objects and functions abstract to morphisms.

Although category theory is a relatively new domain of mathematics, introduced and formulated in 1945 [6], categories are frequently found in this field (sets, vector spaces, groups, and topological spaces all naturally give rise to categories). Category theory offers a number of concepts and theorems about those concepts, which form an abstraction and unification of many concrete concepts in diverse branches of mathematics. Category theory provides a language with a convenient symbolism that allows for the visualization of quite complex facts by means of diagrams.

Compared to other software concept formalizations, category theory is not a semantic domain for formalizing components or their connectors, but expresses the semantics of interconnection, configuration, instantiation, and composition, which are important aspects of modeling the self-* behavior of a complex software system.

4.1 Objects, Morphisms and Social Life

The social life of an object is defined by the interactions of this object with other objects. This interaction in categorical terms is called morphism – a relationship between two objects in a category. Since the interaction is directed, so from graph theory terminology, we can depict the interactions as arrows. A category thus could be seen as a group of objects having a social life defined by arrows between its objects. An absence of arrows means absence of social life. Also, objects could be defined as being of a certain type thus having a certain type of interaction, in turn having a certain type of social life.

In CML notation, let the category be called **MyCategory (C)**. Here, "MyCategory" is the name and "C" is the identifier. All category constructs can be represented visually. Let category 'C' be depicted as a square (see Figure 1). In CML, a category could be simple or typed. For typed category, let the objects in **MyCategory** be assigned a type and based on the type, a name. The type could be defined as: **Obj_TYPE A,** and the names of objects of this type could be defined as A_1, A_2, and A_3. Graphically, the objects are represented using a circle (see Figure 1). Similarly for morphisms, let a type of morphism be defined as **Mor_TYPE: M.** Subsequently, the name of the morphisms under this type could be: m_1, m_2. Notice, the name of the morphisms is in lower-case to avoid confusion with object names. Graphically, the morphisms are depicted as arrows. An example of CML morphism definition is $m_1 : A_5 \rightarrow A_2$ (see Figure 1). The source object (domain) of m_1 is A_5 and its target object (co-domain) is A_2.

4.2 Composition, Associativity and Identity

Like mentioned previously, each object has its own identity. Identity of each object is also represented using arrows. In CML terminology, object A_1 will have identity arrow $Id(A_1)$. Graphically, the identity arrow is a loop arrow (see Figure 1). Usually it is not drawn to avoid cluttering the diagram but is assumed to be present. In a category, two arrows can be composed together, keeping in view the direction of arrows, to make a path. The path could be referred to as the composition morphism or the composition arrow. The composition is represented as '$_o$'. Graphically, the composition arrow is just another arrow with name consisting of the two arrows composed '$m_2 {}_o m_1$' (see Figure 1). The composition arrow $m_2 {}_o m_1$ is read m_1 composed with m_2 keeping in view the direction.

Composing three arrows in a direction also leads to a path and has to evaluate true for associativity. Let's assume we have another object A_4 and three morphisms such that, $m_1 : A_5 \rightarrow A_2$, $m_2 : A_2 \rightarrow A_3$ and $m_4 : A_3 \rightarrow A_1$. So, m_1, m_2, m_4 could be composed to give a path such that $m_4 {}_o (m_2 {}_o m_1) = (m_4 {}_o m_2) {}_o m_1$ (see Figure 1). For every morphism between two objects, e.g. $m_4 : A_3 \rightarrow A_1$, we have composition with the identity arrows of the two objects such that:

$$Id(A_3) {}_o m_2 = m_2 = m_2 {}_o Id(A_2)$$

Graphically this could be represented as shown in Figure 1.

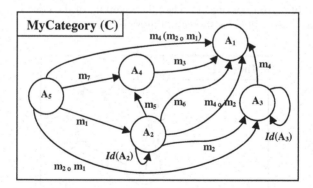

Fig. 1 Graphical Category Showing Objects, Morphisms, Identity, Composition and Associativity.

4.3 Diagram, Cone and Functor

A diagram is a collection of objects and morphisms of a category C. A diagram in a typed category has indexed objects and morphisms. For instance, objects A_1, A_3, A_4 and the morphisms m_3, m_4 form a diagram D in category C shown in Figure 2(b).

In a commutative diagram, any two morphisms with the same domain and co-domain, where at least one of the morphisms is a composition of two or more diagram morphisms, are equal (see Figure 2(b)). The diagram D in category C, Figure 2(b) is extended to a commutative diagram with an *apical* object A_2 and morphisms m_2, m_5 and m_6 with $m_3 \circ m_5 = m_4 \circ m_2$ showing the triangles commute.

The importance of the category theory lies in its ability to formalize the notion that things that differ in substance can have an underlying similarity of "structural" form. A mapping between categories that preserves compositional structure is called a functor. A functor $F: I \rightarrow C$ associates to each object I of I a unique image object $(F(i)$ of C and to each morphism $f : i_1 \rightarrow i_2$ of I a unique morphism $F(f): F(i_1) \rightarrow F(i_2)$ of C, which preserves the compositional structure. An index category provides the required mathematical model for hierarchical structure of knowledge. In our work, an object of an index category represents an abstraction of the type of a real-world entity, the set of input events that activate that type of the entity, and the corresponding output events. An instance of an index category representing a specific configuration of real-world entities is defined as a target category. Let **Index (I)** be the index category, graphically represented in Figure 2(a) and **MyCategory (C)** be the target category, graphically represented in Figure 2(b), where the target category is patterned on the index category. It follows that a functor $F: I \rightarrow C$ maps commutative diagrams of I to commutative diagrams in C. This means that any structural constraints expressed in I are translated into C. For instance, the conical structure shown in Figure 2(b) is called *cone*. Cones for a diagram D in a category are objects of a category **Cone(D)**. In our work, cones will be used to represent the abstraction of the behavior of a group of objects. Then the mapping of the cone in I to a cone in C (see Figure 2(a)) will be as shown in Figure 2(c).

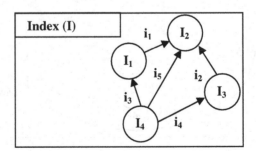

Fig. 2(a) Index Category.

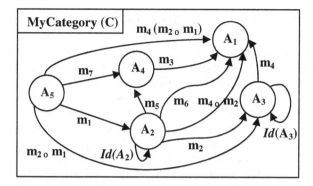

Fig. 2(b) Target Category.

The sole idea is to specify once the structural hierarchy and constraints as an index category, and then (re)define functors mapping the index category to a target category as required by the specific requirements of the modeled software system configuration.

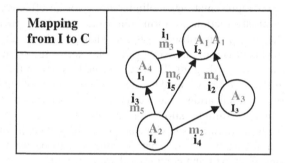

Fig. 2(c) Mapping of Structural Constraints.

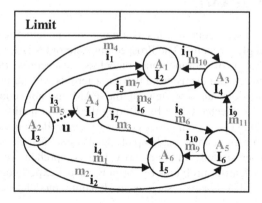

Fig. 2(d) Limit of a Diagram (D) with Objects: I_2, I_4, I_5, I_6.

4.4   Initial and Terminal Objects, Limit and Colimit

Key notions for the theoretical background of this paper are initial and terminal objects in a category. An initial object, where one exists in **C**, is an object for which every other object of **C** is a codomain of a unique morphism with the initial object as a domain (see, for instance, object A_4 in Figure 2(d)). A terminal object has every object of C as the domain of a unique morphism where the terminal object is a codomain. An example of a terminal object is A_2 in Figure 2(d).An important use of these key notions is in the definition of limits and colimits.

Limit is a universal cone over a diagram **D** represented by a terminal object in the category **Cone(D)**. A universal cone will be such that every other cone in the diagram factors through it uniquely (see Figure 2(d)). That is, from every other object in **Cone(D)** there exists a unique morphism '**u**' to the universal cone.

In our modeling approach, limits model the decomposition of high-level view on the complex communications structure into simpler communication views. The limit object I_1 in Figure 2(d) represents an aspect of a group property shared by the objects in the diagram. Dually, Colimit over a diagram **D** is represented by an initial object in the category **Cone(D)**. Colimits model the construction of more abstract representation of communication structure through composition of existing lower level representations. Colimits allow for an effective reuse of the existing behavioral abstraction to form behaviors not yet represented in the categorical model of a given software system. That is, colimits can be used for modeling emergent behavior of an evolving complex software system.

Dually, Colimit over a diagram **D** is represented by an initial object in the category **Cone(D)**. Colimits model the construction of more abstract representation of communication structure through composition of existing lower level representations. Colimits allow for an effective reuse of the existing behavioral abstraction to form behaviors not yet represented in the categorical model of a given software system. That is, colimits can be used for modeling emergent behavior of an evolving complex software system.

5   Case Study (PAM)

Researchers have used agent swarms as a computer modeling technique and as a tool to study complex systems. The Prospecting Asteroid Mission (PAM) involves launching a swarm of autonomous picoclass spacecrafts (approximately 1 kilogram each) to explore the asteroid belt for asteroids with certain characteristics [1]. Figure 3 provides an overview of the PAM mission concept. The PAM spacecrafts study a selected target by offering the highest quality and coverage of measurement through particular classes of measurers, called virtual teams.

A virtual instrument team is made up of members of each class to optimize data collection. Another strategy involves providing comprehensive measurement to solve particular scientific problems by forming virtual experiment teams made up of multiple specialist classes, such as a dynamic modeler team, an asteroid

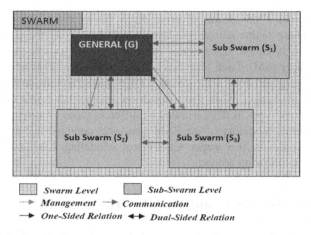

Fig. 3 PAM Mission Concept.

detector and stereo mapper team, a petrologist team, a prospector team, a photogeologist team, etc. The social structure of the PAM swarm can be determined by a particular set of scientific and mission requirements, and representative system elements may include: 1) a General, for distributed intelligence operations, resource management, mission conflict resolution, navigation, mission objectives, and collision avoidance; 2) Rulers, for heuristic operation planning, local conflict resolution, local resource management, scientific discovery data sharing, and task assignment; 3) Workers, for local heuristic operation planning, possible ruler replacement, and scientific data collection [12]. Inside a swarm there are a number of sub-swarms. Figure 3 shows a PAM swarm communication structure.

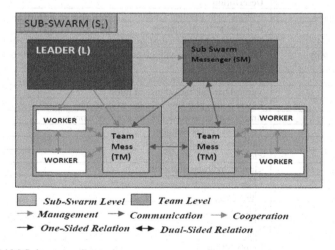

Fig. 4 A PAM Sub-swarm Scenario.

Figure 4 shows a sub-swarm structure. Each sub-swarm consists of teams of spacecrafts. These teams are assigned specific tasks. Inside the sub-swarm the ruler spacecraft is called "Leader (L)". Inside each sub-swarm there is a sub-swarm messenger (SM). Each team has a team messenger (TM) who serves as a communication agent between the 'Worker (W)' spacecrafts inside each team and the sub-warm messenger (SM). The interaction amongst all levels of spacecrafts can be of different types as demonstrated by different types of arrows in Figures 3 and 4. The formal modeling of the PAM structures depicted in Figures 3 and 4 with the proposed in this paper Categorical Modeling Language (CML) is described in the following section.

6 Category Modeling Language (CML)

This section introduces the grammar of CML and describes some examples of categorical representation for the RAS [17] (Reactive Autonomic System) meta-model used for formalizing PAM model. CML uses the Extended Backus-Naur Form (EBNF) for grammar notation to formally specify the syntax. The formal semantics is provided by the meaning of the constructs in the categorical representation. <Category Construct> specifies the subset of categorical constructors applied in this work for modeling complex evolving systems, namely, limit, colimit and types category. In this work we build on the research results on modeling complex software systems such as reactive autonomic systems is described in [16].

Table 1 CML Grammar Conventions

Attribute	Description
<Non-terminal>	Indicates non-terminal symbols
Terminal	Indicates terminal symbols
CONSTRUCT	Terminals in bold face type are reserved words for basic constructs
Construct-Entity	Terminals in bold & italics face are reserve words for parts of a construct
:=	Indicates non-terminal symbol is followed by production rule(s)
|	Vertical bar indicates choice of rules
{ }+	Braces with a plus sign indicates at least once or more
{ }*	Braces with an asterisk indicates zero or more
: ( ) → = ,	Terminals (For separation)
[ ]	Indicates optional expression
<Type>	<Id>
<*_type_Id>	<Id> (* indicates all)
<*_name>	<Id>
<*_type>	<Id>
<*_instance_Id>	<Id>
<*_instance>	<Id>
<*_Id>	<Id>
ε	Empty String

6.1 Grammar Conventions

EBNF grammar consists of "non-terminals" and "terminals." Non-terminals are symbols within a EBNF definition, defined elsewhere in the EBNF grammar. Terminals are endpoints in an EBNF definition, consisting of category theoretic keywords. In this section, all non-terminals appear in brackets (< >) and all terminals appear without brackets.

6.2 Category Theory Grammar

The grammar has been defined for almost all of the constructs of category theory. The category has been defined as Typed by default. Since, almost every object and morphisms will have a type describing the structural application of the construct to different case studies.

```
<Category Construct> :=   <Typed_Category> | <Category> | <Sub_Category> | <Functor> |
                          <Typed_Functor> | <N_Trans> | <Diagram> | <Cone> | <Co-Cone> |
                          <Limit> | <Co-Limit> | <Product> | <Co-Product> | <Push-Out> |
                          <Pull-Back> | <Slice> | <Co-Slice>
```

Fig. 5 Category Theory Grammar

Typed_Category. This paper proposes a principle of typed-category construct providing a mechanism for standardization of hierarchical structure and behavior of the colonies of interacting objects as well as properties (laws) that must restrict the emergent behavior and system evolution. Typed-categories require the existence of index categories defining the types of objects and morphisms, target categories capturing the instances of those object types and morphism types, and a

```
<Typed_Category> := TYPED-CATEGORY
                    <Cat_name> (<Cat_Id>)
                    Types of Objects
                    {<Object_Type>}+
                    Types of Morphisms
                    {<Morphism_Type>}+
                    Morphism Instances
                    {<Morphisms>}+
                    Axioms <Axioms>
<Object_Type>:=     Obj_Type : <Obj_type>(<Obj_type_Id>)
                    [Obj_Sub-type: <Obj_type>(<Obj_type_Id>) {, Obj_type>(<Obj_type_Id>) }* ]
<Morphism_Type>:=   Mor_Type: <Mor_type> (<Mor_type_Id>)
                    Mor_Sub-type: <Arrows>
<Arrows>:=          <Obj_type_Id> → <Obj_type_Id> | <Obj_type_Id> → <Obj_type_Id>,<Arrows>
<Arrow_Identity>:=  Identity
                    {Id(<Obj_type_Id> <Obj_instance_Id>):
                    <Obj_type_Id><Obj_instance_Id> → <Obj_type_Id><Obj_instance_Id>, }+
<Axioms>:=          {<Associativity>}*
<Associativity>:=   Associativity:
                    <Mor_type_Id> <Mor_instance_Id> o <Mor_type_Id> <Mor_instance_Id> o
                    <Mor_type_Id> <Mor_instance_Id>) = (<Mor_type_Id> <Mor_instance_Id> o
                    <Mor_type_Id> <Mor_instance_Id>) o <Mor_type_Id> <Mor_instance_Id>,
<Id>:=                   <Character><Id> | <Empty>
<Character>:=       A | B | C | ... |Z | a | b | c | ... | z | 0 | 1 | 2 | ... | 9
<Empty>:=           ε
```

Fig. 6 Grammar for Typed Category Construct

functor (homomorphical mapping) between index and target categories. Types and properties give strong guidance to the developer of the specified complex software system in what it accomplishes, without impinging on the how, which further refines the compositional style of categorical language semantics. The idea is simply that each instance of a index category, that is, the target category, follows the semantics abstracted in the index category and hence builds in correctness by construction. It also serves to transfer laws about a type's semantic model. The grammar for typed-category is given in Figure 6.

There should be at least one object in the typed-category and identity morphisms for all objects. Objects, morphisms both have types. The "axiom" symbol is used to define both associativity and identity composition.

Typed_Functor. The grammar for typed-functor is based on the functor construct of category theory. The source category is mapped to the target category. Both source and target category symbols are symbols with typed-category grammar. Object and morphism mapping symbols hold the grammar for the respective mapping.

```
<Typed_Functor> := TYPED-FUNCTOR
                  (<Functor_name>(<Functor_Id>))
                  Source Typed Category
                  <Typed_Category>(<Cat_name>,<Cat_Id>)
                  Target Types Category
                  <Source_Typed_Category> (<Cat_name>,<Cat_Id>)
                  Object Mapping
                  {<ObjectsM>}*
                  Morphism Mapping
                  {<MorphismsM>}*
<ObjectsM>:=      <Obj_type_Id> :
                  (<Obj_Subtype_Id>:<Obj_instance_Id>; {, <Obj_instance_Id>}*);
                  ({, <Obj_Subtype_Id>:<Obj_instance_Id> {, <Obj_instance_Id>}* }* );
<MorphismsM>:=    <ArrowsM>
<ArrowsM>:=       <Mor_type_Id> : (<Obj_type_Id> → <Obj_type_Id> : <Mor_instance_Id>); |
                  <Mor_type_Id>;
                  (<Obj_type_Id> → <Obj_type_Id> : <Mor_instance_Id> , <ArrowsM>);
```

Fig. 7 Grammar for Typed Functor Construct

6.3 PAM Sub-swarm Modeled Using Typed Category and Typed Functor

The sub-swarm scenario discussed in section 5 could be modeled using the Typed-Category grammar. Figure 8 represents the PAM Sub-swarm structure. The model demonstrates the semantics' expression and abstraction flexibility of category theory.

Figure 8 models part of a basic PAM sub-swarm. The model shown in Figure 9 represents a part of a typical PAM sub-swarm structure.

The otherwise complex architecture has been simplified into a category of interacting objects. The CML model for the PAM sub-swarm given in Figure 8 has captured the work structure of a sub-swarm using typed-category and the semantics of the structure in the form of the typed-functor. The same constructs could be used to model the structure of other parts of a particular PAM sub-swarm

TYPED-CATEGORY
Sub-Swarm (S$_1$)
Types of Objects
Obj_Type: Ruler (R)
Obj_Sub-Type: Leader (L)
Obj_Type: Messenger (M)
Obj_Sub-Type: Team Messenger (TM), Sub-Swarm Messenger (SM)
Obj_Type: Worker (W)
Obj_Sub-type: Imaging (W$_{IM}$), Altitude(W$_{AL}$), Infra-Red(W$_{IR}$)
Type of Morphisms
Mor_Type: Management (m)
Mor_Sub-type: L → TM, L → SM, L → W
Mor_Type: Cooperation (c)
Mor_Sub-type : W → W, TM → W
Mor_Type: Communication (cu)
Mor_Sub-type : TM → SM, TM → TM
Morphism Instances
m$_1$: L$_1$ → TM$_1$, m$_2$: L$_1$ → SM$_1$, m$_3$: L$_1$ → W$_{IM3}$, c$_1$: W$_{IM1}$ → W$_{AL}$, c$_2$:W$_{IM1}$ → W$_{IR1}$, c$_3$: W$_{IR1}$ → W$_{AL}$,
c$_4$: TM$_2$ → W$_{IM1}$, c$_5$:TM$_2$ → W$_{AL}$, c$_6$:TM$_2$ → W$_{IR1}$, c$_7$:TM$_1$ → W$_{IM3}$,
cu$_1$:TM$_1$ → SM$_1$, cu$_2$: TM$_2$ → SM$_1$, cu$_3$: TM$_2$ → TM$_1$
Identity
Id(L$_1$): L$_1$ → L$_1$, Id(SM$_1$): SM$_1$ → SM$_1$, Id(TM$_1$): TM$_1$ → TM$_1$, Id(TM$_2$): TM$_2$ → TM$_2$,
Id(W$_{IM1}$): W$_{IM1}$ → W$_{IM1}$, Id(W$_{IM3}$): W$_{IM3}$ → W$_{IM3}$, Id(W$_{IR1}$): W$_{IR1}$ → W$_{IR1}$, Id(W$_{AL}$): W$_{AL}$ → W$_{AL}$
Axioms
Associativity: c$_3$ o (c$_2$ o c$_4$) = (c$_3$ o c$_2$) o c$_4$

TYPED-FUNCTOR
TypeMap (TM)
Source Typed Category
Sub-Swarm (SW)
Target Typed Category
SW (S$_1$)
Object Mapping
R:(L:L$_1$);
M:(TM:TM$_1$, TM$_2$), (SM: SM$_1$);
W:(W$_{IM}$: W$_{IM1}$, W$_{IM3}$), (W$_{AL}$, W$_{AL1}$), (W$_{IR}$: W$_{IR1}$);
Morphism Mapping
m:(L → TM: m$_1$), (L → SM: m$_2$), (L → W: m$_3$);
c:(W → W: c$_1$), (TM → W: c$_2$);
cu:(TM → SM: cu$_1$), (TM → TM: cu$_2$);

Fig. 8 CML Model for a PAM Sub-swarm Scenario

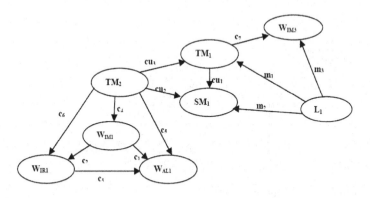

Fig. 9 Graphical PAM Sub-swarm Category

or any sub-swarm for the matter. The abstract nature of category, objects and morphisms makes it possible to model and visualize architecture, semantics and behaviour using the same constructs. The structure for example could be in one category and the semantics in another and the structure could now be mapped to the semantics using the functor construct.

Diagram. A diagram is a group of objects and their related morphisms inside a category. This construct provides the flexibility of grouping components. This would further aid in modeling the behaviour of a group of objects inside a category of objects. The grammar for this construct is as follows.

```
<Diagram>:=      Index Category : <Typed_Category>
                 Target Category : <Typed_Category>
                 DIAGRAM
                 (<Diag_Id>,<Cat_Id>,<Cat_Id>)
                   Diagram Objects
                   {<D_Objects >}⁺
                   Diagram Morphisms
                   {<D_Morphisms>}⁺
<D_Objects>:=        Obj_Type: <Obj_type>; <Obj_indexing> {, <Obj_indexing>}*
<Obj_indexing>:=    <Diag_Id> ( <Vertex_index_Id>) : <Cat_Id> ( <Vertex_index_Id>) →
                       <Cat_Id> (<Diag_Id> (<Obj_type_Id><Obj_instance_Id>))
<D_Morphisms>:=   Mor_Type: <Mor_type>: <Mor_indexing> { , <Mor_indexing>}*
<Mor_indexing>:=  <Diag_Id> (<Edge_index_Id>) : <Cat_Id> (<Vertex_index_Id>,
                       <Vertex_index_Id>, <Edge_index_Id> →   <Cat_Id> (<Diag_Id>
                       (<Obj_type_Id><Obj_instance_Id>), <Diag_Id><Obj_type_Id><Obj_instance_Id>) )
```

Fig. 10 Grammar for Diagram Construct

Limit. Limit of a certain group of objects or a diagram to be precise is a specialized cone as seen previously in section 4 where the cone consists of a diagram and a specialized object named *apical* object. This apical object in our PAM example is assigned a Ruler role. The grammar for Limit is given in Figure 11.

```
<Limit>:=  Category: <Typed_Category> (<Cat_name> , <Cat_Id>)
           Diagram: <Diagram> (<Diag_name><Diag_Id>)
           Cones
           {(<Obj_type_Id>,<Obj_instance_Id>)}⁺
             LIMIT
             (Universal Cone: <Universal_Cone>,
             Unique Morphism: <Mor_Unique>)
             Limit Objects
             {<L_Objects>}⁺
             Limit Morphisms
             {<L_Morphisms>}⁺
             Limit Axioms
             {<L_Axiom>}⁺
             <Universal_Cone>:= <Obj_type_Id><Obj_instance_Id>
             <Mor_Unique>:= <Mor_Id>: <Cone_Obj> (<Vertex_index_Id>) → <Universal_Cone>
             <L_Objects>:= <Diag_Id> (Vertex_index) {, <Diag_Id> (Vertex_index)}*
             <L_Morphisms>:= Mor_Type: <Mor_type> : <Cone_Obj> (<Vertex_index_Id>) :
                       <Cone_Obj> → <Diag_Id>(<Vertex_index_Id>)
                       {, <Cone_Obj> (<Vertex_index_Id>) : <Cone_Obj> →
                       <Diag_Id>(<Vertex_index_Id>)}* | Mor_Type: <Mor_type> :
                       <Diag_Id> (<Edge_instance_Id>) : <Diag_Id> (<Vertex_index_Id>) →
                       <Diag_Id> (<Vertex_index_Id>) {, <Diag_Id> (<Edge_instance_Id>) :
                       <Diag_Id> (<Vertex_index_Id>) → <Diag_Id> (<Vertex_index_Id>)}*
             <L_Axiom>:=   <Mor_Unique> o <Cone_Obj>(<Vertex_index_Id>)
                       = <Cone_Obj>(<Vertex_index_Id>)
```

Fig. 11 Grammar for Limit Construct

Figure 12 includes Petrologist Team structure diagram (**D**). The indexes are marked in bold. A Leader in a sub-swarm guides the rest of the spacecrafts in that sub-swarm. This concept could be modeled using the Limit construct in Category Theory. Figure 12 shows *'Limit'* with a unique morphism 'u' coming from the apical object TM4 of the corresponding cone to the apical object L2 of the Limit

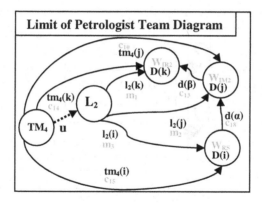

Fig. 12 Graphical Limit Cone for Petrologist Team Diagram

cone. Limit defines the category-wide behavior of a group of objects (diagram) with certain specialized objects (Apical object). This category-wide behavior may or may not exist. But the Cone construct would still model the local behavior of the group of spacecrafts inside a sub-swarm with a 'Ruler' object on top.

6.4 PAM Petrologist Team Messenger Modeled Using LIMIT Construct

The Petrologist Team structure modeled using the grammar given in Figures 9 and 10 is illustrated in Figure 13. The graphical counterpart of the CML model aids in visualizing the structure effectively representing the behaviour of this sub-swarm and the social life of its objects.

Category :(Petrologist Team Category, PGT)
Diagram :(Petrologist Team Diagram, D)
Cones (TM$_4$, L$_2$)
LIMIT
Universal Cone: L$_2$
Unique Morphisms: u: TM$_4$→ L$_2$
Limit Objects
d(i), d(j), d(k)
Limit Morphisms
Cooperation: tm$_4$(k): TM$_4$ → D(k), tm$_4$(j): TM$_4$ → D(j), tm$_4$(i): TM$_4$ → D(i),
                  l$_2$(k): L$_2$ → D(k), l$_2$(j): L$_2$ → D(j), l$_2$(i): L$_2$ → D(i)
Cooperation: d(α): D(i) → D(j), d(β): D(j) → D(k)
Limit Axioms
u o l$_2$(i) = tm$_4$(i), u o l$_2$(j) = tm$_4$(j), u o l$_2$(k) = tm$_4$(k)

Fig. 13 Graphical Limit Cone for Petrologist Team Diagram

7 Category Theory Modeler (CATOM)

CATOM serves as a graphical modeler tool for the visualization and formalization of complex software systems based on category theory construction. CATOM is modeler tool that lets you draw and edit models for systems based on frameworks

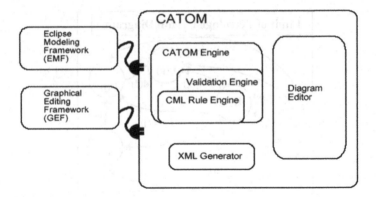

Fig. 14 CATOM Architecture

like (Reactive Autonomic Systems Framework (RASF) [17] and SWARM [1] using different constructions from category theory as the underlying logic. Figure 14 shows the architecture of the tool CATOM. The model is drawn using the Diagram Editor in collaboration with the CATOM Engine. The CATOM engine consists of a validation module and a CML rule engine module the work in collaboration to assist semantically correct model. The CML model is saved in XML format. For the current release of CATOM, the diagram/model is drawn using the editor which is validated by the CATOM Validation feature.

CATOM is currently going through an extension and revision phase. The tool will allow it's users to describe, analyze and reason about a model of a system using CML.

8 Conclusion and Future Work

Category theory provides an approach for modelling a complex system at different levels of abstraction as demonstrated in the PAM case-study. The structure of a system with interacting and cooperating objects with or without semantics' representation could be easily visualized when modeled using tool in support of the CML. The flexibility of the same constructs to model different levels of complexity (e.g. architecture, behavior) is one of the few arguments that advocate for the use of category theory for making sense out of the complexity in a software system. The future work to this report includes modelling of the PAM cross-swarm and sub-swarm interaction in terms of behaviour, synchronization and cooperation using other CML constructs.

References

1. Hinchey, M.G., Sterritt, R., Rouff, C.: Swarms and Swarm Intelligence. IEEE, Los Alamitos (2007)
2. Whitmire, S.: Object Oriented Design Measurement. Whiley C. P, New York (1997)
3. Barr, M., Wells, C.: Category Theory for Computing Science. Prentice-Hall, NJ (1990)

4. Fiadeiro, J.: Categories for Software Engineering. Springer, Heidelberg (2005)
5. Goguen, J.A.: A categorical manifesto. Mathematical Structures in Computer Science 1, 49–67 (1991)
6. Eilenberg, S., Mac Lane, S.: General Theory of Natural Equivalences. Transactions of the American Mathematical Society 58, 231–294 (1945)
7. Pfalzgraf, J.: ACCAT tutorial. Presented at 27th German Conference on Artificial Intelligence (2004)
8. Lee, W. M.: Modelling and Specification of Autonomous Systems using Category Theory. PhD Thesis, University College of London, London, UK (1989)
9. Hinchey, M.G., Rouff, C.A., Rash, J.L., Truszkowski, W.F.: Requirements of an Integrated Formal Method for Intelligent Swarms. In: Proceedings of the 10th International Workshop on Formal Methods for Industrial Critical Systems, Lisbon, Portugal, pp. 125–133 (2005)
10. Vassev, E., Hinchey, M., Paquet, J.: A Self-Scheduling Model for NASA Swarm-Based Exploration Missions Using ASSL. In: Proceedings of the 5th IEEE Workshop on Engineering of Autonomic and Autonomous Systems, Belfast, Northern Ireland, pp. 54–64 (2008)
11. Truszkowski, W.F., Hinchey, M.G., Rash, J.L., Rouff, C.A.: Autonomous and Autonomic Systems: a Paradigm for Future Space Exploration Missions. IEEE Transaction on Systems, Man, and Cybernetics, Part C: Applications and Reviews 36(3), 279–291 (2006)
12. Curtis, S., Mica, J., Nuth, J., Marr, G., Rilee, M., Bhat, M.: ANTS (Autonomous Nano Technology Swarm): an Artificial Intelligence Approach to Asteroid Belt Resource Exploration. In: Proceedings of the 51st International Astronautical Congress (2000) IAA-00-IAA.Q.5.08
13. Wiels, V., Easterbrook, S.: Management of Evolving Specifications Using Category Theory. In: Proceedings of the 13th IEEE International Conference on Automated Software Engineering, pp. 12–21 (1998)
14. Pfalzgraf, J.: On an Idea for Constructing Multiagent Systems (MAS) Scenarios. In: Advances in Multiagent Systems, Robotics and Cybernetics: Theory and Practice, IIAS, Tecumseh, ON, Canada, vol. 1 (2006)
15. Fiadeiro, J.L., Maibaum, T.: A Mathematical Toolbox for the Software Architect. In: Proceedings of the 8th International Workshop on Software Specification and Design, Schloss Velen, Germany, pp. 46–55 (1996)
16. Kuang, H., Ormandjieva, O., Klasa, S., Khurshid, N., Benthar, J.: Towards Specifying Reactive Autonomic Systems by a Categorical Approach. In: Proceedings of International Conference on Software Engineering Research, Management and Applications (SERA 2009), China (2009)
17. Ormandjieva, O., Kuang, H., Vassev, E.: Reliability Self-Assessment in Reactive Autonomic Systems: Autonomic System-Time Reactive Model Approach. ITSSA 2(1), 99–104 (2006)
18. Harel, D.: Statecharts: A visual formalism for complex systems. Science of Computer Programming 8, 231–274 (1987)
19. Harel, D.: On Visual Formalisms. ACM 31, 514–530 (1988)
20. Selic, B.: Using UML for Modeling Complex Real-Time Systems. In: Müller, F., Bestavros, A. (eds.) LCTES 1998. LNCS, vol. 1474, pp. 250–260. Springer, Heidelberg (1998)

A Survey on the Importance of Some Economic Factors in the Adoption of Open Source Software

Vieri Del Bianco, Luigi Lavazza, Sandro Morasca, Davide Taibi, and Davide Tosi

Abstract. Economic advantages have long been used as a key factor for promoting the adoption of Open Source Software. This paper reports on an investigation about the impact of economic factors when deciding on the adoption of Open Source Software, in the framework of a survey carried out in the QualiPSo project. The results seem to indicate that economic issues may have a remarkably lower impact than commonly believed, though people with roles more directly related to economic results and working in private companies seem to give economic factors more consideration than other Open Source Software stakeholders.

1 Introduction

The usage of Open Source Software (OSS) has been continuously increasing in the last few years, mostly because of the success of a number of well-known projects. OSS products are characterized by the free availability of the source code and a number of other software artifacts. In addition, the development process of OSS is "open" too [11]. Contributions to the source base are often made by volunteers and there is no rigidly organized development process. OSS can be used directly by end users or it can be customized, modified, or integrated into other products. So, a number of different categories of OSS stakeholders exist: developers, integrators, system administrators, product managers, clearing house members, end users, etc.

Given its success, OSS is no longer viewed as an amateur kind of software development. A number of major international software players have become more and

Vieri Del Bianco
Systems Research Group, CASL, University College Dublin, Dublin, Ireland
e-mail: vieri.delbianco@ucd.ie

Luigi Lavazza, Sandro Morasca, Davide Taibi, and Davide Tosi
Dipartimento di Informatica e Comunicazione,
Università degli Studi dell'Insubria, Como, Italy
e-mail: {luigi.lavazza,sandro.morasca,davide.taibi}@uninsubria.it
davide.tosi@uninsubria.it

Roger Lee (Ed.): SERA 2010, SCI 296, pp. 151–162, 2010.
springerlink.com   © Springer-Verlag Berlin Heidelberg 2010

more interested in developing and using OSS, and they have created internal OSS development and assessment units, because they see OSS as a good opportunity to do business.

Along with ethical and technical ones, financial reasons have often been touted as very important in the adoption of OSS. The free availability of the source code has often been mentioned as one of the key factors in the introduction of OSS in industrial environment (even though it has been pointed out that OSS is not necessarily free of charge [4, 3]). The idea, at any rate, is that the adoption of OSS can basically provide value for free, to all of the above mentioned OSS stakeholders. For instance, end users may view OSS as a software product they can use to improve their own productivity; developers and integrators as a way to reduce effort and time needed to build a software product, while keeping quality at a high level and receiving support from the OSS development community; software managers as a way to reduce costs, keep schedules, and successfully deliver software products for specific customers or the general marketplace. Note that the benefits of the openness of OSS span beyond the mere financial factors. For example, end users and all stakeholders in general may benefit from the knowledge of the list of known faults.

At any rate, it has been long known that, when carrying out a fair cost-benefit analysis of OSS, it is necessary to look beyond the fact that an OSS may be free of charge. The cost of OSS, like the cost of many other products, needs to be evaluated along all of its lifecycle. One needs to ascertain or at least estimate the various kinds of costs that may be encountered during and following the adoption of OSS. OSS stakeholders may not be totally aware of these costs. For example, the customization of an OSS product to fit the specific needs of an organization may very well not be free of charge. Charging for customization is the basis of the business model of a number of software developers that use OSS as their primary software source. This business model has several analogies, like, for instance, in the case of computer printers. The computer printer itself may be bought for a relatively low price, but then the ink cartridges needed to actually utilize the printer may be quite expensive. Also, there may be costs related to switching to OSS from a previously used product. Users or developers may need to learn the features and/or structure of the OSS product they are adopting, with at least temporary loss of productivity.

All of these costs may be viewed as investments, which are supposed to yield a return at some point in the future. So, the Return on Investment (ROI) and the Total Cost of Ownership (TCO) are usually believed to be two important factors that need to be taken into account when deciding whether to adopt a specific OSS product.

Financial advantages and value, however, are not the only things that matter when adopting OSS, like any other product. A number of other concerns come into play. For instance, the possibility of receiving support, should any problems arise, is often considered a real issue in the OSS world. In this respect, OSS still bears the business stigma of its early days, when it was mostly developed by volunteer enthusiasts, who would not have any responsibilities in case of problems, nor would be legally or otherwise bound to provide any kind of support for bug fixing or adding new functionalities or perfecting and customizing existing ones. OSS stakeholders would know of no clear roadmap for the development of an OSS product, so for instance there would be no assurance that the new version of an OSS product would still be compatible with the old ones. Even more importantly, OSS stakeholders would have no assurance that support may be available for a given OSS product, even not for free. OSS development communities may start and end without any notice, and there would be no recourse for OSS stakeholders. Thus, OSS stakeholders need to trust OSS software. As the example about support shows, the notion of trustworthiness is not simply confined to technical issues, e.g., the quality or the security of and OSS product. An OSS product is trustworthy if itself and its development and maintenance process can be trusted. This also has an impact on ROI and TCO. An OSS that is untrustworthy is likely to have lower ROI and higher TCO, as it will likely imply higher costs, which, in addition, may not be easily estimated beforehand.

This paper reports on a survey that has been carried out in the framework of QualiPSo [10], an EU-funded research project on the trustworthiness of OSS. Our results provide evidence that may be somewhat contrary to conventional wisdom. OSS stakeholders as a whole do not seem to attach a high degree of importance to ROI or TCO when deciding on the adoption of an OSS product. However, OSS stakeholders that have roles more directly related to economic results or who work in private companies may give economic factors more consideration than other OSS stakeholders. The interested reader should refer to [10] for more information on 1) the study whose part related to economic factors is summarized in this paper, 2) all of the investigations carried out about the trustworthiness of OSS products, and 3) the QualiPSo project in general.

The remainder of this paper is organized as follows. Section 2 describes the structure of the questionnaire that was used to elicit information rom the interviewees. Section 3 describes how the survey was carried out. Section 4 describes information about the sample (Section 4.1), the data analysis we carried out on the economic factors we investigated (Section 4.2), and additional information we obtained in our interviews (Section 4.3). Section 5 discusses some relevant threats to the validity of the study. Conclusions and an outline for future work are in Section 6.

2 The Questionnaire

With our questionnaire, we attempted to obtain answers on a number of factors that may be believed to affect OSS stakeholders' decisions when deciding whether to adopt an OSS product or component. The questionnaire was also developed keeping into account the actual literature on software product quality and OSS trustworthiness (e.g., [1, 5, 6]).

The questionnaire we developed is a general-purpose one: it can be used if OSS is used as is or is developed/modified; it is applicable to companies of any size; it targets any role (from the inexperienced developer to upper management levels); and it is not related to a specific application domain. In addition, we endeavor to understand how OSS is perceived by people from different types of ICT companies and with different roles.

The questions in the questionnaire can be mainly classified in three different categories:

1. *Organization, project, and role.* These questions are needed for profiling the interviewed person, the company he or she works for, the project(s) he or she participates in. Some of this information is obviously private and is collected for profiling reasons only. During the survey, it was made clear to the interviewees that this information would not be disclosed at all, and that, in the presentation of results, all information would be disclosed in aggregated form, so as to make it impossible to identify single respondents or single companies and their answers. Section 4.1 contains more detail on the profiling information we collected.
2. *Actual problems, actual trustworthiness evaluation processes, and factors.* These questions are needed to identify the main factors actually considered when evaluating whether to adopt an OSS product.
3. *Wishes.* These open-answer questions are needed to understand what should be available but is not, and what indicators should be provided for an OSS project to help its adoption. So, these questions addressed possible wishes that OSS stakeholders have on the information they would like to have about OSS products.

We grouped the factors in the following categories

- *Economic.* We asked about the importance of ROI and TCO, and any other indicators that may be important (these would be mentioned by the interviewees themselves in the "wishes" open questions).
- *Development.* For instance, we asked about the importance of the specific license, the availability of short-term support, the availability of documentation, etc., by means of closed- and open-answer questions. So, this category also included process factors that may influence OSS adoption decisions.
- *Product quality.* In this category, we included closed- and open-answer questions about the importance of external and internal qualities [9] of OSS. For instance, we asked about the importance of functionality, reliability, and maintainability as for external qualities, and the importance of modularity, structural complexity, and size, as for internal software qualities.
- *Customer.* The questions in this categories investigated the importance of customer-related issues, such as customer satisfaction and the existence of standards the software used by the customers needed to abide to.

In several questions, we asked the interviewees to provide an indication of the importance they give to each factor when they adopt OSS products. This importance was measured on a 0 to 10 scale, with value 0 meaning "not important at all" and value 10 meaning "of fundamental importance." The idea was not to actually attach absolutely precise meanings to these numbers, but to provide interviewees with a way to give us their idea of the relative importance ordering among these factors.

We began with a list of 35 factors. However, two more factors were included in our analysis as a result of the open questions, in which the interviewees could mention additional factors that they deemed important which we had not included in our initial list of factors.

3 The Empirical Study

We carried out a total of 151 interviews with respondents from Italy, Germany, France, Spain, Poland, Brazil, China, United Kingdom, and USA. The vast majority of the interviews were carried out in person and a few by phone. We believe this is the most effective way to elicit information and establish an effective communication channel with the interviewees. We wanted to collect information that was structured by means of closed-answer questions and additional information with open-answer questions and by talking with the interviewee.

We also carried out interviews by email, giving feedback and advice in an asynchronous way. The results seem to be fairly aligned and coherent with the direct interviews, but of poorer quality. For instance, we obtained far fewer details on open-answer questions. When the differences between the questionnaire obtained in a synchronous way and the questionnaires obtained in an asynchronous way become clear, we decided to continue with interviews only in person or by telephone.

All the interviews we carried out were individual ones, usually with one interviewee at a time, since we believed that it is important that the interviewees provide their own viewpoint without any sort of conscious or even unconscious interference due to the presence of other people, especially if belonging to the same organization.

4 Results

We first provide information about the sample of respondents (Section 4.1), which can be used to better interpret the statistical analysis results we discuss in Section 4.2. We also provide some additional information on economic issues that we collected from open-answer questions in Section 4.3. Even though it has not been subject to quantitative analysis, this information may provide useful knowledge about the eoconomic factors used when adopting OSS.

4.1 The Sample

We now provide information about a few relevant characteristics of the sample of interviewees. Table 1 contains the percentages of the roles for four organizational roles. Note that roles may not necessarily be mutually exclusive.

Table 1 Organizational Roles of Respondents

Role	%Yes	%No
Upper Management	30.8%	69.2%
Project Manager	20.5%	79.5%
Developer	39.7%	60.3%
OSS Expert	6.4%	93.6%

Table 2 reports on the education of the respondents. Note that the education degrees are not mutually exclusive. For instance, all PhD respondents also possessed some other education degree. Also, different educational systems exist in our respondents' countries, so we grouped them in common categories. For instance, "College 2-3 years" approximately groups the respondents that achieved a degree similar to an undergraduate degree. "College 4-5 years" groups those respondents that achieved a degree that required 4 or 5 years without any intermediate degree.

Table 2 Education of Respondents

Degree	%Yes	%No
High School	94.9%	5.1%
College 2-3 years	67.9%	32.1%
College 4-5 years	51.3%	48.7%
Master	38.5%	41.5%
PhD	11.5%	88.5%

Table 3 contains the percentages about the role of OSS in the respondents' organizations. Note that the education degrees are not mutually exclusive.

Table 3 Role of OSS in Respondents' Organizations

Role	%Yes	%No
Support software development	69.2%	30.8%
Part of other products	64.1%	35.9%
Customized or configured	79.5%	20.5%
Support for internal processes	65.4%	34.6%
Provide services to the outside world	66.7%	33.3%
Development platform	73.4%	26.6%
Target usage platform	76.1%	23.9%

Finally, Table 4 reports on the types of the organizations our respondents belonged to.

Table 4 Type of Organization

Type	Public	Private	No profit
Percentage	10.4%	80.5%	9.1%

4.2 Data Analysis

The data analysis we carried out allowed us to establish statistical significant ranking relationships between the 37 factors. We could safely carry out such an analysis because the values provided by the respondents are on an ordinal scale. We used the Sign Test, the Mann-Whitney Test, and the Wilcoxon Test [7] to assess the statistical significance of the relative rankings. As is usual in Empirical Software Engineering, we used a 0.05 statistical significance threshold.

The statistical analysis has actually allowed us to partition the factors in 8 separate groups, with the most important factors in group 8 and the least important ones in group 1. Our statistical analysis has also provided evidence for the existence of an ordering between factors belonging to different groups. For instance, customer-related factor "customer satisfaction" turned out to belong to group 7 and therefore it was believed to be more important than product-related factor "modularity," which belongs to group 6. No ordering can be established among the factors belonging to the same group. For instance, we do not have statistically significant supporting evidence to say that product-related factor reliability is more important than product-related factor maintainability or *vice versa*, since both are in group 7.

We also computed the means of the values provided by the respondents, because this is an expressive piece of information, as the 11-valued scale we used may be considered a Likert scale. At any rate, we found an almost perfect concordance between the ranking group of a factor and the mean value we obtained. By considering all of the 666 pairs of factors, the ordering between two factors according to their group differ from the ordering between them according to the mean value of respondents only for 12 pairs.

Not unexpectedly, the fact that an OSS product or component satisfies the functional requirements needed by the OSS stakeholder and its reliability appeared to be the most important factors. Somewhat surprisingly, structural size came in last, as the least important factor. This is somewhat surprising if one considers that a number of Empirical Software Engineering prediction models are primarily based on size (e.g., COCOMO [2]).

In general, both ROI and TCO were expected to be considered very important, but the results do not support this intuition. Table 5 contains the results we obtained on ROI and TCO, where

- "mean" is the arithmetic mean of the values provided by the interviewees
- "more" is the number of factors that turned out to be more important than the factor
- "equal" is the number of factors that turned out to be neither more nor less important than the factor
- "less" is the number of factors that turned out to be less important than the factor.

These unexpected results could be partly explained by the type of organization, education, and organizational role of the interviewees, as shown in Table 6, where

- "condition" shows the value of some profiling information on interviewees

Table 5 Ranking of ROI and TCO

Factor	Group	Mean	More	Equal	Less
ROI	3	5.722	22	6	8
TCO	2	5.633	29	5	2

- "mean if true" shows the value of the arithmetic mean obtained only with those interviewees whose profiling information is reported in column "condition"
- "mean if false" shows the value of the arithmetic mean obtained only with those interviewees whose profiling information is different from that reported in column "condition."

For instance, the average value provided by respondents in 'No profit" organizations as to the importance of ROI as a factor for adopting OSS is 3.167, while the mean value provided by all other respondents is 6.629. Specifically, the table only reports on those factors in which the impact of the truth value of a condition on the mean obtained is statistically significant at the 0.05 statistical significance level. For instance, the mean value provided by respondents in 'No profit" organizations as to the importance of ROI as a factor for adopting OSS is statistically significantly different from the mean value provided by all other respondents. We did not report on non statistically significant differences between means.

Table 6 Impact of Organization, Education, and Organizational Role on Means

Condition	Factor	Mean if true	Mean if false
type of organization: No profit	ROI	3.167	6.629
type of organization: Private	ROI	6.891	3.923
education: master	ROI	5.167	7.000
education: phd	ROI	3.750	6.705
education: phd	TCO	3.667	6.415
org role support: internal processes	TCO	5.531	7.160
org role support: sw development	TCO	5.558	7.318

4.3 Information from Open-Answer Questions

Other economic related factors and issues have been mentioned as important by at least some of the respondents as a part of the open questions. Here, we report a summary of the issues collected.

- *Ethics*. OSS experts and OSS supporters support ethic values instead of direct economical profits.
  - *Social cost*. Social cost may be considered as important as direct cost; this factor can be related to the more general ethics factor.

- *Development time.* Delivery time may be held as more important than the total cost of the product.
- *No use.* A very small number of the interviewees' organizations do not use OSS products a priori.
- *Closed specifications.* In some organizations, software systems are developed to fulfill closed specifications, which cannot be freely distributed. Hence, the implementation of closed specifications in a software product to be distributed sometimes negates the possibility to use OSS products.
- *Integration cost and effort.* Some products need to be integrated with existing software. Integration cost and effort have been reported to be high if there is the need to integrate proprietary software with OSS.
- *Risk analysis.* Risk analysis may be very important when evaluating the acquisition and use of a software system.
- *OSS market.* The possibility of becoming the driving force behind some OSS market niche by developing OSS is considered as the one of the driving economic factor.
- *Differentiate from competitors.* OSS software can be a distinguishing factor when compared to competitive products. This factor is going to lose some of its strength once OSS software is more widely adopted.
- *Full control of code.* This is considered an important economic factor, since unwanted economic dependencies can be avoided.

  - *Ability to contribute to evolve and adapt the software.* This factor can be considered a sub factor of full control of code.
  - *Independence from specific vendors and commercial products.* Independence and no vendor lock-ins are very important economic dependencies to be avoided. This factor can be considered a sub factor of full control of code.

- *ROI*

  - *Absence of license fees.* This factor stresses the fact that software licenses will be acquired for free, hence increasing the ROI.
  - *Try many solutions without spending money.*

- *TCO*

  - *Preference to stay with the same OSS product because expertise was acquired, and this reduces the effort.* This factor can be seen as a characteristic of TCO.

- *Acquisition*

  - *Ease of acquisition.* Ease of acquisition, especially for support and assistance services of OSS products, is considered important.
  - *Rules for spending money.* In many organizations, spending money to buy software can be a lengthy and complicate process. Since there is usually no money to be spent at the moment of OSS acquisition, OSS is regarded as a faster and easier way to acquire the needed software.

5   Threats to Validity

A number of threats may exist to the validity of a study like ours. We now examine some of the most relevant ones.

5.1   Internal Validity

Consistent with the literature, we used a 0.05 statistical significance threshold for the rankings and the impact of interviewees' characteristics of Section 4. We used non parametric statistical tests like the Sign Test, the Mann-Whitney Test, and the Wilcoxon Test [7], which are appropriate for ordinal variables like ours. We also dealt with the impact of interviewees' characteristics by taking the means of the values provided by the interviewees. This may yield results that are not always fully significant from a practical point of view, since comparing mean values of ordinal variables may not be significant from a strict Measurement Theory point of view [12, 8]. However, we believe that we have at least some evidence about the impact of those factors on the responses.

5.2   External Validity

The threats to the external validity of our study need to be identified and assessed. The most important issue is about the fact that our sample may not be fully "balanced," and that may have somewhat influenced the results. While this may be true, the following points need to be taken into account.

- It was not possible to interview several additional people that could have made our sample more "balanced," because they were not available or had no or little interest in answering our questionnaire.
- No reliable demographic information about the overall population of OSS stakeholders is available, so it would be impossible to know if a sample is "balanced" in any way.
- Like in many other empirical studies, we used a so-called "convenience sample," composed of respondents who agreed to answer our questions. We collected information about the respondents' experience, application field, etc., but we did not make any screening beforehand. Excluding respondents based on some criteria, which must have been perforce subjective, may have resulted in an "unbalanced" sample, which may have biased the results.
- We used the profiling information to find out which of the interviewees' characteristics may have an impact on the responses, as we showed in Section 4.
- We dealt with motivated interviewees, so this ensured a good level for the quality of responses.
- There is no researcher's bias in our survey, since we simply wanted to collect and analyze data from the field, and not provide evidence supporting or refuting some theory.

5.3   Construct Validity

An additional threat concerns the fact that the measures used to quantify the relevant factors may not be adequate. Our rankings are based on a 0 to 10 scale which allowed us to have sufficient variation to rank the factors and have indications on the impact of interviewees' characteristics on their responses. The very nature of our study required that we collect subjective data, as we wanted to capture OSS stakeholders' opinions, so no "objective" measure could be used to collect that information anyway.

6   Conclusions and Future Work

Our results, obtained in a survey that has been carried out in the framework of QualiPSo [10], an EU-funded research project on the trustworthiness of OSS, provide evidence somewhat contrary to conventional wisdom. OSS stakeholders as a whole do not seem to give a high degree of importance to ROI or TCO when deciding on the adoption of an OSS product. However, OSS stakeholders that have roles more directly related to economic results or who work in private companies give economic factors more consideration than other OSS stakeholders.

Future work will include carrying out additional interviews to widen the statistical sample we have and study if additional profiling factors of the respondents may be related to the adoption of OSS.

Acknowledgements. The research presented in this paper has been partially funded by the IST project "QualiPSo," funded by the EU in the 6th FP; the FIRB project "ARTDECO," sponsored by the Italian Ministry of Education and University; and the projects "Elementi metodologici per la descrizione e lo sviluppo di sistemi software basati su modelli" and "La qualità nello sviluppo software," funded by the Università degli Studi dell'Insubria. We also would like to thank all of the interviewees that participated in the survey.

References

1. Antikainen, M., Aaltonen, T., Väisänen, J.: The role of trust in oss communities – case linux kernel community. In: Feller, J., Fitzgerald, B., Scacchi, W., Sillitti, A. (eds.) OSS. IFIP, vol. 234, pp. 223–228. Springer, Heidelberg (2007)
2. Boehm, B.W., Clark, B., Horowitz, E., Westland, J.C., Madachy, R.J., Selby, R.W.: Cost models for future software life cycle processes: Cocomo 2.0. Ann. Software Eng. 1, 57–94 (1995)
3. Cerri, D., Fuggetta, A.: Open standards, open formats, and open source. The Journal of Systems and Software 80, 1930–1937 (2007)
4. Fuggetta, A.: Open source software – an evaluation. The Journal of Systems and Software 66, 77–90 (2003)
5. Hansen, M., Köhntopp, K., Pfitzmann, A.: The open source approach opportunities and limitations with respect to security and privacy. Computers & Security 21, 461–471 (2002)

6. ISO/IEC, ISO/IEC TR 9126 parts 1-4 (2001-2004)
7. Kanji, G.: 100 statistical tests. Sage, London (1994)
8. Morasca, S.: On the definition and use of aggregate indices for nominal, ordinal, and other scales. In: IEEE METRICS: Proceedings of the 2004 International Symposium on Software Metrics, pp. 46–57. IEEE Computer Society, Washington (2004)
9. Morasca, S.: A probability-based approach for measuring external attributes of software artifacts. In: ESEM 2009: Proceedings of the 2009 3rd International Symposium on Empirical Software Engineering and Measurement, pp. 44–55. IEEE Computer Society, Washington (2009)
10. Trust and Quality in Open Source Systems (QualiPSo), http://www.qualipso.eu
11. Raymond, E.S.: The Cathedral and the Bazaar. O'Reilly & Associates, Sebastopol (1999)
12. Roberts, F.: Measurement theory, with applications to Decision Making, Utility and the Social Sciences. Addison-Wesley, Boston (1979)

Verification of the Correctness in Composed UML Behavioural Diagrams

Samir Ouchani, Otmane Ait Mohamed,
Mourad Debbabi, and Makan Pourzandi

Abstract. The Unified Modeling Language UML 2.0 plays a central role in modern software engineering, and it is considered as the de facto standard for modeling software architectures and designs. Today?s systems are becoming more and more complex, and very difficult to deal with. The main difficulty arises from the different ways in modelling each component and the way they interact with each others. At this level of software modeling, providing methods and tools that allow early detection of errors is mandatory. In this paper, a verification methodology of a composition of UML behavioural diagrams (State Machine, Activity Diagram, and Sequence Diagram) is proposed. Our main contribution is the systematic construction of a semantic model based on a novel composition operator. This operator provides an elegant way to define the combination of different kind of UML diagrams. In addition, this operator posses a nice property which allows to handle the verification of large system efficiently. To demonstrate the effectiveness of our approach, a case study is presented.

Keywords: Transition System, Unified Modelling Language (UML), Model Checking, Security Properties.

1 Introduction

A major challenge in the software development process is to advance error detection to early phases of the software life-cycle. For this purpose, the verification of UML diagrams plays an important role in detecting flaws at the design level. It has a distinct importance for software security, since it is crucial to detect security flaws

Samir Ouchani, Otmane Ait Mohamed, and Mourad Debbabi
Concordia University, Montreal, Canada
e-mail: {s_oucha,ait,debbabi}@ece.concordia.ca

Makan Pourzandi
Ericsson Software Research, Montreal, Canada
e-mail: makan.pourzandi@ericsson.com

Roger Lee (Ed.): SERA 2010, SCI 296, pp. 163–177, 2010.

before they have been exploited. ¿From the literature, a lot of techniques have been proposed for verification of softwares as well as hardwares like: Model Checking, Theorem Proving, and Static Analysis, etc. The most of the techniques used for verification of UML diagrams is model checking. Model checking is an important technology of automatic verification. It verifies the properties against a model through explicit state exploration, and elegantly presents the counter example paths when the system does not satisfy a property. The most important researches focusing on model checking of UML models verify the properties specified by a formal language after extracting the semantic model of UML design, and then they translate it into the input languages of the existing model checkers.

Most of the approaches proposed in the literature are intended either to activity, state machine [1, 3, 5, 6, 8, 9, 12, 14], or sequence diagrams [1, 13] separately. We experience in industrial collaborations, that in practice most UML behavioural diagrams are mixed and connected. Effectively, in the literature there is a poor prior approachs that proposes a solution for this case when UML behavioural diagrams interacts.

The main intent of our work is to focus on the verification of UML design models containing different connected UML behavioral models. We will focus on security properties like authentication. For that we have chosen the model checking technique because it's automatic, and characterized by features like model reduction. To construct the semantic model of different UML behavioural diagrams, we defined a new compositional operator to fully automate the semantic model generation of the interacted UML models. The security properties are specified by a simple instantiation from security templates describing a set of application-independent properties to produce a set of application-dependent properties proper to the application[5].

As a case study, we apply the proposed technique to verify the message authentication security property on Automated Teller Machine (ATM). ATM is written as an UML-based model composed of two different UML behavioural diagrams:a state machine describing client authentication and an activity diagram describing transaction operation. The ATM security properties were obtained from the authentication templates, and formalized by the formal language : the Computational Tree Logic (CTL)[2]. The result of this case study shows how to verify a complex system described by a mixed UML behavioural diagram.

The remainder of this paper is organized as follows: Section 2 presents the related work. Section 3 explores UML behavioral diagrams (BDs) and their possible interactions inside a UML design. We define and generate the semantic model for the global UML design in the form of transition system in Section 4. The proposed verification approach is detailed in Section 5. In Section 6, we provide a ATM case study. Finally, we conclude the paper by a conclusion and a promising future work in Section 7.

2   Related Work

In the state of the art, there is a considerable number of researches which are intended to verify just one kind of UML behavioural bediagrams like a state machine or a sequence diagram without taking in consideration their interactions when a state machine call an activity diagram for example.

Cheng et al. [5] investigate how the verification of security properties can be enabled by adding formal constraints to UML-based security patterns. ¿From the proposed templates, they instantiate the security properties to enable their analysis by using the Spin model checker. The limit of their work is in how to tailor a security pattern to meet the needs of a system especially when it contains a set of connected diagrams.

A Static Verification Framework (SVF) is developed in [14] to support the design and verification of secure peer-to-peer applications. The framework supports the specification, modeling, and analysis of security properties together with the behavior of the system. The SVF developped in [12] is the continuation work of Andrea [14], they translate the UML state machine including guards into Promela models that are amenable for Spin model checking with LTL properties. Their works are limited to only the state machine diagram that describe a peer-to-peer applications, also their properties represents just a proposed scenarios for the attacker.

Beato et al. [3] developed a complete automatic tool called TABU to transform active and state machine diagrams into an SMV (Symbolic Model Verifier) specification via Labeled Transition Systems (LTS). The properties are specified by the pattern classification proposed by Dwyer et al [7]. In their work, they didn't consider when a state machines are composed with activity diagrams.

Eshuis et al. describe in [8] a tool that verify UML activity diagrams by specifying the activity diagrams as a Clocked Transition Systems then generate automatically the NuSMV input code. The limits of their approach resides when they ignore the verification of some cases by stopping the computation of the transition system if one of the nodes becomes unbounded, and more than that, they focuse only on activity diagram.

Giese et al. provide in [9] a domain specific formal semantic definition to verify a real-time UML design and an integrated sequence of design steps by prescribing how to compose complex software systems from domain-specific patterns. The composition of these patterns to describe the complete component behaviour is prescribed by a syntactic definition which guarantees the verification of components and system behaviour can exploit the results of the verification of individual patterns. Amstel et al.[13] propose trace analysis techniques by using model checkers to improve the quality of sequence diagrams, and to get PROMELA code from sequence diagrams. This technique provides a translation scheme that is defined in [11]. These works are limited to just one kind of real-time diagram as well as sequence diagram.

Dong et al. in [6], define a set of rules to verify UML Dynamic Models. These rules are based on hierarchical automata between semantics structures, and simulation relation to reduce the detailed components to an abstracted specifications. In their work, they gave the results without linking their simple case study to the theory proposed. More than that, they didn't mention the effect of the simulation relation in a model which is the main step in verification.

In [1], a framework has been proposed for verifying UML diagrams, the extracted semantics model is called Configuration Transition System (CTS), a kind of Transition System. The resulting CTS is translated into NuSMV [10] code. This approach allows verfication behavioral against properties written in CTL, and our work is the continuation of [1].

The verification in [1, 3, 5, 6, 8, 9, 12, 13, 14] is done by using different semantic models, and using different checking techniques. None of them addressed the problem of linking several UML Behavioural Diagrams.

3   Syntax of UML Behavioral Diagrams

UML supports behavioral modeling, but more than that it supports the interaction between behavioral models. Figure 1 shows the different UML behavioral models. In this section, we explore all the possible interactions between UML behavioral diagrams to complete the systax defined in [1].

3.1   State Machines (SM)

In UML, we have two kinds of State Machines (SM): *behavioral state machines* and *protocol state machines*. The role of these two diagrams is to express the behavior of the system, and the usage protocol of part of that system, respectively. A SM can have association with a BD in its state as in its transition:

- State: It models a situation where some invariant condition holds. A state can have one association with a BD in three places:

  1. *doActivity*, a BD is executed while being in the state. The execution starts when this state is entered, and stops either by itself or when the state is exited whichever comes first.

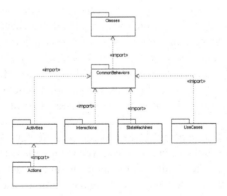

Fig. 1 all the possible interactions between UML behavioral diagrams.

2. *Entry*, a BD is executed whenever the state is entered regardless of the transition taken to reach the state. If defined, entry actions are always executed to completion prior to any internal behavior or transitions performed within the state.
3. *Exit*, a BD is executed whenever this state is exited regardless of which transition was taken out of the state. If defined, exit actions are always executed to completion only after all internal activities and transition actions have completed execution.

- Transition: It is a directed relationship between a source vertex and a target vertex in a state machine. During the activation of this transition, the behavioral diagram specified in association with *Effect* can be executed.

3.2   Activity Diagram (AD)

An activity diagram includes concurrent control, data flow, and decisions. It supports structured activities (sequences, loops, and conditions), but it can have just one association with another behavior at time for each of the following activity elements:

- DecisionNode: It is a control node that chooses between outgoing flows. It has an association in *decisionInput* to provide input to guard specifications on edges outgoing from the decision node.
- ObjectFlow: It is an activity edge to model the flow of values to/or from object nodes that can have objects or data passing along them. It can have two associations with a BD in *Selection* object to select tokens from a source object node, and in *Transformation* object to change or replace data tokens flowing along edge.
- ObjectNode: It's an abstract activity node that contains only values at runtime that conform to the type of the object node. It has an association with a BD in *Selection* to select tokens for outgoing edges.

3.3   Interaction Diagram(ID)

Interactions are a mechanism for describing systems that can be understood and produced by providing different capabilities that makes it more appropriate for certain situations (i.e, Sequence Diagrams, Interaction Overview Diagrams, Communication Diagrams, Timing Diagrams and Interaction Tables). This kind of diagrams can have association with only one BD at time in a *BehaviorExecutionSpecification* element. It is a kind of *ExecutionSpecification* representing the execution of a behavior. A *BehaviorExecutionSpecification* can be associated with those BD having their execution occurring in a Behavior element.

For the remainder of this paper we note S_{elmt} as the set of elements where the BDs can be connected.

4 Semantic of UML Behavioral Diagrams

4.1 Configuration Transition System

A Configuration Transition System (CTS)[1] is a formal description of the behavior of a system. A CTS is considered as a directed graph where nodes represent *configurations*, and edges model *transitions*. A *Configuration* is a specific binding of a set of values to the set of variables in the dynamic domain of the behavior of a system. i.e, the set of value assigned to the variables of the system during one step of execution (e.g., the evaluation of variables in an iteration of a program). A *Transition* specifies how the system can change from one configuration to another i.e, the relation between the current configuration and next ones. Definition 1 [1] gives the formal definition of a CTS.

Definition 1 (Configuration Transition System). A Configuration Transition System CTS is a tuple $(C, Act, \rightarrow, I, E)$ where:

- C is a set of configurations,
- Act is a set of actions,
- $\rightarrow \subseteq C \times Act \times C$ is a transition relation,
- I is the initial state,
- E is the final state.

Large systems are built from smaller parts, so we have to reason about their components, and how they interacts. Semantically, a component Π_i of a system is represented by CTS_i and the parallel composition of the system $\Pi_1 \parallel \ldots \parallel \Pi_n$ is represented by the composition of CTSs components $CTS_1 \circ \ldots \circ CTS_n$. In our case, the composition is defined as a simple substitution (*Definition 2*), we substitute the transition relation that represents an element of S_{elmt} (called *interface*) by its corresponding CTS.

Definition 2 (Composition of CTS). Let CTS_i, and CTS_j be two CTS where $CTS_i = (C_i, Act_i, \rightarrow_i, I_i, E_i)$, and $CTS_j = (C_j, Act_j, \rightarrow_j, I_j, E_j)$.
   The composition $(CTS_i \circ CTS_j)$ of CTS_i, and CTS_j is the substitution of CTS_j in the specific transition that represents the interface $\rightarrow_r (c_{i1}, act_i, c_{i2})$ of CTS_i defined by the tuple: $CTS = (S, Act, \rightarrow, I, E)$, where:

- $C = (C_i \cup C_j) \setminus \{I_j, E_j\}$,
- $Act = (Act_i \cup Act_j) \setminus \{act_i\}$,
- $\rightarrow \subseteq (\rightarrow_i \cup \rightarrow_j) \setminus \rightarrow_r$,
- $I = I_i$,
- $I_j = c_{i1}, E_j = c_{i2}$, and $E = E_i$.

In order to ensure the scalability of the verification process for systems composed as *Definition 1* we have derived *Theorem 1*.

Theorem 1. *The composition of CTS's is associative, i.e.:* $(CTS_i \circ CTS_j) \circ CTS_k = CTS_i \circ (CTS_j \circ CTS_k)$.

Proof. Consider three configuration transition systems CTS_i, CTS_j, and CTS_k. CTS_i compose CTS_j in the transition (called the interface) $\rightarrow_r (c_{i1}, act_i, c_{i2})$, and CTS_j compose CTS_k in a the transition $\rightarrow_l (c_{j1}, act_j, c_{j2})$.
$CTS_i \circ CTS_j = \Gamma (C_1, Act_1, \rightarrow_1, I_1, E_1)$ where:

- $C_1 = (C_i \cup C_j) \setminus \{I_j, E_j\}$,
- $Act_1 = (Act_i \cup Act_j) \setminus \{act_i\}$,
- $\rightarrow_1 \subseteq (\rightarrow_i \cup \rightarrow_j) \setminus \rightarrow_r$,
- $I_1 = I_i$,
- $I_j = c_{i1}$, $E_j = c_{i2}$, and $E_1 = E_i$.

And $CTS_j \circ CTS_k = \Lambda (C_2, Act_2, \rightarrow_2, I_2, E_2)$ where:

- $C_2 = (C_j \cup C_k) \setminus \{I_k, E_k\}$,
- $Act_2 = (Act_j \cup Act_k) \setminus \{act_j\}$,
- $\rightarrow_2 \subseteq (\rightarrow_j \cup \rightarrow_k) \setminus \rightarrow_l$,
- $I_2 = I_j$,
- $I_k = c_{j1}$, $E_k = c_{j2}$, and $E_2 = E_j$.

From another side we have $\Sigma' = \Gamma \circ CTS_k$ in $\rightarrow_l (c_{j1}, act_j, c_{j2})$ and $\Sigma'' = CTS_i \circ \Lambda$ in $\rightarrow_r (c_{i1}, act_i, c_{i2})$, and by the same composition we will have $\Sigma' \left(C', Act', \rightarrow', I', E' \right)$ where:

- $C' = (C_1 \cup C_k) \setminus \{I_k, E_k\}$,
- $Act' = (Act_1 \cup Act_k) \setminus \{act_j\}$,
- $\rightarrow' \subseteq (\rightarrow_1 \cup \rightarrow_k) \setminus \rightarrow_l$,
- $I' = I_i$,
- $I_k = c_{j1}$, $E_k = c_{j2}$, and $E' = E_i$.

and $\Sigma'' \left(C'', Act'', \rightarrow'', I'', E'' \right)$ where:

- $C'' = (C_i \cup C_2) \setminus \{I_2, E_2\}$,
- $Act'' = (Act_i \cup Act_2) \setminus \{act_i\}$,
- $\rightarrow'' \subseteq (\rightarrow_i \cup \rightarrow_2) \setminus \rightarrow_r$,
- $I'' = I_i$,
- $I_2 = c_{i1}$, $E_2 = c_{i2}$, and $E'' = E_i$.

From the two previous result we find: $\Sigma' \equiv \Sigma''$, so the composition is associative.

More than that, we conclude from *Definition 2* the maximum number of possibilties to apply the compositional operation in the initial CTS_1. This maximum is bounded by the number of interfaces and its up to the number of transitions (n). Also, we can observe that this composition is not commutative, and not transitive.

4.2 Generation of Configuration Transition System

To capture the semantic model of a single BD having no interaction with other BDs, we have to generate its proper CTS where each configuration represents the active

---

Algorithm 1. The CTS of a simple diagram

```
SingleCTS (FoundConfList ,CTSConfList, CTSTransList, EventList:list)
begin
    while FoundConfList IsNotEmpty do
        CurrentConf = pop(FoundConfList);
        if CurrentConf not in CTSConfList then
            | CTSConfList = CTSConfList ∪ {CurrentConf};
        end
        NextTransList=getNext(CurrentConf,EventList);
        for nextTrans in NextTransList do
            | nextConf = getDestination(nextTrans);
            | FoundConfList = FoundConfList ∪ { nextConf };
        end
        CTSTransList = CTSTransList ∪ NextTransList;
    end
end
```

Algorithm 2. The CTS of a composition of diagrams

```
ComposedACTS (d: diagram, FoundConfList ,CTSConfList, CTSTransList,
EventList:list)
begin
    while FoundConfList IsNotEmpty do
        CurrentConf = pop(FoundConfList);
        if CurrentConf not in CTSConfList then
            | CTSConfList = CTSConfList ∪ { CurrentConf };
        end
        NextTransList=getNext(CurrentConf,EventList);
        for nextTrans in NextTransList do
            if nextTrans HasInterface then
                if d not computed then
                    | CTS(FoundConfList,CTSConfList,
                    | CTSTransList,EventList,d);
                end
            end
            nextConf = getDestination(nextTrans) FoundConfList = FoundConfList ∪
            { nextConf };
        end
        CTSTransList = CTSTransList ∪ NextTransList;
    end
end
```

---

elements in that diagram, and the transition represent the transition from source configuration to the target configuration. To achieve that, we iterate the breadth-first search procedure as presented in *Algorithm 1* which is the simplified version of the one presented in [1]. In each iteration, the new configuration explored from the current configuration denoted by *CurrentConf* and the trace of configurations are saved in *FoundConfList* list where *CTSTransList* list contains the transitions between configurations. The unexplored configurations are in *CTSConfList* list, and *EventList* lists the possible incoming events. Initially, still they are a discovered configurations to explore in *FoundConfList*, the top element is loaded into *CurrentConf* and add it into result list of configuration if it's not added in *CTSConfList*. Given the current configuration *CurrentConf* and the event list *EventList*, the trace of configuration *FoundConfList* is updated. While this trace is not empty the configuration transition list *CTSTransList* will be updated.

To generate the CTS of a UML diagram interconnected with other BDs by association presented in section 2, and noted by $\rightarrow_r$ in definition 2 we propose the recursive algorithm *Algorithm 2* derived from *Algorithm 1*. The algorithm calls itself when a transition contains an interface. However, we avoid to recalculate the CTSs that have been already generated so far.

5 Verification Methodology

Our contribution is an automatic model checking based approach as depicted in Figure 2. The approach consists of two parts: *the verification part* where we construct the global semantic model of our design model, and the second one is *the Specification part* where we express a set of security properties to be verified for our model. At the beginning of *the verification part*, we have a set of separated UML behavioral diagrams, for this reason we have to extract their corresponding semantic models (CTS) separately, and based on their interactions in the global model we construct its corresponding CTS by constructing the set of the existing interfaces between the CTSs of the small parts. The formal structure CTS representing the global system to be verified is a composition of a small ones: $CTS_1, \ldots, CTS_n$ where $CTS = CTS_1 \circ \ldots \circ CTS_n$. Our objectif is to check for a given property (π) if it holds for CTS verifying $CTS \models \pi$ (i.e. $CTS_1 \circ \ldots \circ CTS_n \models \pi$) involves the exhaustive inspection of CTS instead of the property π. The property we want to verify should be formally specified by a büchi automata, or using a temporal logic to be able to use model checking. Based on *Definition 2*, and *Theorem1* we have derived a corollary to handle the verification of the complex system having n sub-components.

Corollary 1. *Let* $CTS = CTS_1 \circ \ldots \circ CTS_i \circ \ldots \circ CTS_n$ *be a CTS composed of n-sub-CTS, and* π *a property, the following expression is always true:*
$[(CTS_i \circ \ldots \circ CTS_n) \models \pi] \Rightarrow [CTS \models \pi] (1 \leq i < n).$

Proof. Here we will use the same formula and indications used in the previous proof to prove the first part.

Let's for a given property (π), and a global model $CTS = CTS_1 \circ \ldots \circ CTS_n$. Here we like to prove the following expression: $[(CTS_i \circ \ldots \circ CTS_n) \models \pi] \Rightarrow [CTS \models \pi]$
From *the theorem1*, we can write $CTS = \Gamma \circ \Lambda$ where: $\Gamma = CTS_1 \circ \ldots CTS_{i-1}$, and, $\Lambda = CTS_i \circ \ldots \circ CTS_n$. So we have to prove the following: $(\Lambda \models \pi) \Rightarrow [(\Gamma \circ \Lambda) \models \pi]$.
Consider π a sequence of states: $\pi = (s_0, \ldots, s_n)$
$(\Lambda \models \pi) \Leftrightarrow (\exists \hookrightarrow \subseteq \rightarrow_2 : \hookrightarrow \equiv \pi).$

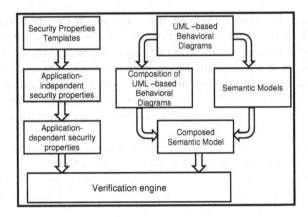

Fig. 2 Compositional Verification Approach.

From the previous proof, we have $\rightarrow_2 \subseteq \rightarrow$ So, $\pi \subseteq \rightarrow$ which mean $(\Gamma \circ \Lambda) \models \pi$. Finaly, $CTS \models \pi$.

From *Corollary 1*, we notice that in order to verify a property we don't need to construct the whole semantic model of the system but only a subset. The main advantage of the *Corollary 1* is to accelerate and optimize the verification procedure.

The specification part: Our approach is a model checking based, so we have chosen to formalize our security properties by using a temporal language like (LTL, CTL, CTL*)[4] depending on the model checker used. For that, we are using a set of security templates proposed in [5]. Firstly, we extract the application-independent security properties from a given templates of security patterns like: *Single Access Point, Check Point, Roles, Session, Full View with Errors, Limited View, Authorization, Multi-level Security*. Secondly, we instantiate their appropriate application-dependent security properties and formalize them using temporal language of the adopted model checker such as Computational Tree Logic (CTL) for NUSMV[1] model checker [10] in our case. From the semantic model defined in section 1 we generate the appropriate NuSMV code to be inputs to NuSMV model checker with the CTL expression of the instantiated security properties.

As an example of this kind of specification, we picked the template of message authentication presented in Figure 3 as a UML state machine where a countermeasure is taken if an authentication is failed. From this template, we can extract the following security properties, and we express them by a macro language, to be applied in the case study section.

1. An unauthorized access leads eventually to the activation of a countermeasure. The corresponding property expressed in the macro language is the following:
   Always (UnautorizedAccess *imply eventually* (CounterMeasureTaken))

---

[1] http://nusmv.irst.itc.it

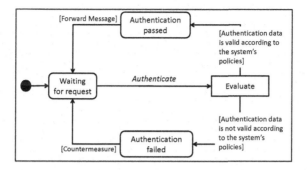

Fig. 3 UML state machine template for the message authentication property.

2. When a request was denied, it is important that the current request remains unsuccessful until a new request is received. The corresponding property expressed in the macro language is the following:
   Always ((UnautorizedAccess *imply* Request Unseccessful) *until* NextRequest)
3. When an access is granted, then it should eventually be able to access the system successfully until the next request is received. The corresponding property expressed in the macro language is the following:
   Always ((AccessGranted *imply eventually* Operation) *until* NextRequest)

Using the Graphviz [2] drawing tool, the CTS corresponding to the verified system, as well as the NuSMV assessment results (i.e., the counterexample), will be visualized graphically.

6 Case Study

An Automated Teller Machine (ATM) is a system that interacts with a potential customer (user) via a specific interface and communicates with the bank over an appropriate communication link.

A user that requests a service from the ATM has to insert an ATM card and enter a personal identification number (PIN). Both information need to be sent to the bank for validation, if the credentials of the customer are not valid, the card will be ejected out. These tasks of validation and ejecting are presented by an UML state machine diagram with *effect* (see Section 3) as described in Figure 4. Otherwise, the customer will be able to perform one or more transactions where the card stays retained in the machine during the customer interaction until the customer wishes no further service. This part of transaction is described by the activity diagram depicted in Figure 5.

To assess the composed diagram showed in Figure 4, we compose its CTS semantic models corresponding to state machine in Figure 6 and activity diagram in Figure 7 with the transition representing the interface of the effect as shown in

---
[2] www.graphviz.org

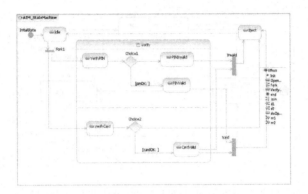

Fig. 4 A state machine specifying the behavior of an ATM.

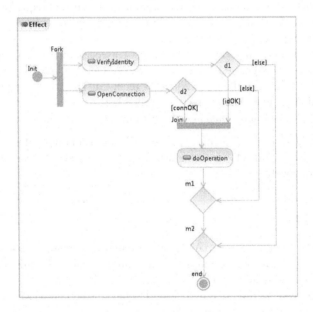

Fig. 5 specifying the behavior of an ATM transaction.

Figure 8. From the message authentication template and their associated application-independent properties defined in Section 5, we instantiate their corresponding ATM dependent properties formulated with CTL as follow:

1. A wrong Pin or Card leads automatically to ejecting the card.
   CTL: $AG\left((!CardValid|!PINValid)\right) \rightarrow EF\left(Eject\right)$
2. A wrong Pin or Card remains an unsuccessful connection till a new request.
   CTL: $A[EF[(!CardValid|!PINInvalid) \rightarrow AX!OpenConnection]UIdle]$

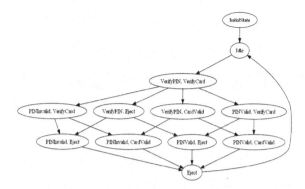

Fig. 6 The CTS corresponding to the State Machine of Figure 4.

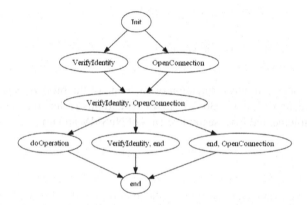

Fig. 7 The CTS corresponding to the Activity Diagram of Figure 5.

3. When both Card and Pin are valid, then it should eventually be able to make an operation until the next request is received.

CTL: $A[AG(CardValid \& PINValid) \rightarrow A[A[!OpenConnection\ U\ VerifyIdentity]\ U\ Eject]$

The operators used in the above properties are a mix of logical (!:Not, |:Or, &: And,→: Imply), and temporal operators (A: All, E: Exists, X: neXt, G: Globaly, F: Finally, U: Until).

To take advantage of *the Corollary 1*, we verify the following property:

4. $AG(Init \rightarrow EF(OpenConnection \rightarrow EF(doOperation)))$

So without constructing the whole CTS, just by verifying the property in the CTS of the activity diagram we conclude that the first property is validated for the whole model.

The verification of the above properties are done by the model checker NuSMV version 2.4.3 reveals the validation of the two first properties, and the fourth one. The violation of the third property as showed by the counterexample in Figure 8. The system diameter(minimum number of iterations of the NuSMV model to obtain all

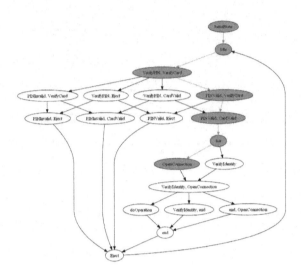

Fig. 8 A counterexample for the third property

the reachable states) in the verification of each one of the properties is 10, and the number of reachable states is 36 out of 245760 and this is due to cone of influence algorithm implemented from second version of NuSMV and up.

7 Conclusion

In this paper we extended the formal verification from individual UML behavioural diagrams into more complex interactions between state machine, activity, and interaction UML behavioural diagrams. This approach allows verification engineer to detect flaws in the earlier stage of software cycle for more wide and complex systems. In fact this is what happens especially in industry where different diagrams interact together. In addition, our approach gives flexibility to write very expressive properties while hiding temporal operator. This technique, along with other verification tools, can provide a powerful and very useful framework to detect errors at the design phase, resulting in reliable software at the end of the software development process.

As a future work, we target two main problems. Firstly, we intend to improve the verification approach in order to deal with more very large and more complex system by splitting the property and distributing sub-properties to the affected parts of the system to the end to achieve a conccurent verification, and develop a formalism proof rules to guaranty the satisfaction of the main property for the whole system. Secondly, The verification of the interacted diagrams must be done without changing their semantic models and verification tools if exists in the separate case in term to produce an optimal verification (cost/size of states). Even in this context, it's very interesting to develop these perspective theory and applications for the industry as acadimia challenges.

References

1. Alawneh, L., Debbabi, M., Jarraya, Y., Soeanu, A., Hassayne, F.: A unified approach for verification and validation of systems and software engineering models. In: ECBS 2006: Proceedings of the 13th Annual IEEE Interntl. Symp. and Works. on Eng. of Comp. Based Sys., pp. 409–418. IEEE Computer Society Press, Washington (2006)
2. Baier, C., Katoen, J.P.: Principles of Model Checking. MIT Press, New York (2008)
3. Beato, M.E., Barrio-Solrzano, M., Cuesta, C.E., de la Fuente, P.: Uml automatic verification tool with formal methods. Electronic Notes in Theoretical Computer Science 127(4), 3–16 (2005); Proceedings of the Workshop on Visual Languages and Formal Methods (VLFM 2004)
4. Bérard, B., Bidoit, M., Finkel, A., Laroussinie, F., Petit, A., Petrucci, L., Schnoebelen, P.: Systems and Software Verification. In: Model-Checking Techniques and Tools. Springer, Heidelberg (2001)
5. Cheng, B.H.C., Konrad, S., Campbell, L.A., Wassermann, R.: Using security patterns to model and analyze security. In: IEEE Workshop on Requirements for High Assurance Systems, pp. 13–22 (2003)
6. Dong, W., Wang, J., Qi, Z., Rong, N.: Compositional verification of uml dynamic models. In: APSEC 2007: Proceedings of the 14th Asia-Pacific Soft. Eng. Conf., pp. 286–293. IEEE Computer Society Press, Washington (2007)
7. Dwyer, M.B., Avrunin, G.S., Corbett, J.C.: Patterns in property specifications for finite-state verification. In: ICSE 1999: Proc. of the 21st Internatnl Conf. on SE, pp. 411–420. ACM Press, New York (1999)
8. Rik, E., Roel, W.: Tool support for verifying uml activity diagrams. IEEE Transactions on Software Engineering 30 (2004)
9. Giese, H., Tichy, M., Burmester, S., Flake, S.: Towards the compositional verification of real-time uml designs. SIGSOFT Softw. Eng. Notes 28(5), 38–47 (2003)
10. Giunchiglia, C.C., Cimatti, A., Clarke, E., Giunchiglia, F., Roveri, M.: Nusmv: a new symbolic model verifier, pp. 495–499. Springer, Heidelberg (1999)
11. Leue, S., Ladkin, P.B.: Implementing and verifying msc specifications using promela/xspin. In: Proceedings of the DIMACS Workshop SPIN 1996, pp. 65–89 (1997)
12. Siveroni, I., Zisman, A., Spanoudakis, G.: Property specification and static verification of uml models. In: ARES 2008: Proceedings of the 2008 Third Interntl Conf. on Avail., Reliab. and Sec., pp. 96–103. IEEE Computer Society Press, Washington (2008)
13. Van Amstel, M.F., Lange, C.F.J., Chaudron, M.R.V.: Four automated approaches to analyze the quality of uml sequence diagrams. In: COMPSAC 2007: Proceedings of the 31st Annual International Computer Software and Applications Conference, pp. 415–424. IEEE Computer Society Press, Washington (2007)
14. Zisman, A.: A static verification framework for secure peer-to-peer applications. In: ICIW 2007: Proceed. of the 2nd Internatnl Conf. on Internet and Web Applic. and Serv., IEEE Computer Society, Washington (2007)

References

Development of Mobile Agent on CBD

Haeng-Kon Kim and Sun Myung Hwang

Abstract. Mobile device has been considered a key technology for embedded software and ubiquitous era. Because, existing web environments is
moving to wireless internet, the new concepts for wireless internet computing environments has gained increasing interest. Mobile agent provide a new
abstraction for deploying functional over the existing infrastructures. Mobile
application systems requires the flexibility, adaptability, extensibility, and autonomous. A main nature of ad hoc mobile networks is frequent change on
their topology that is the source of many problems to be solved. AODV is an
on-demand routing protocol for decreasing maintenance overhead on ad hoc
networks. But some path breaks can cause significant overhead and transmission delays. If the maintenance overhead of routing table can be reduced,
table-driven routing methods could be an efficient substitution. In this paper,
we propose a knowledge discovery agent for an effective routing method that
is using the simple bit-map topology information. The agent node gathers
topology knowledge and creates topology bit-map information. All paths for
source to destination can easily be calculated by the bit-map. All the other
nodes on the network maintain the bit-map distributed from agent and uses
it for source of routing. Correctness and performance of the proposed agent
method is verified by computer simulations

Keywords: Component Based Development, Mobile Agent, Agent Classification, Knowledge Discovery Agent, Ad hoc networks.

Haeng-Kon Kim
Department of Computer information & Communication Engineering,
Catholic Univ. of Daegu, Korea
e-mail: `hangkon@cu.ac.kr`

Sun Myung Hwang
Department of Computer Engineering, Daejeon University,
Dae Jeon, 300-716, Korea
e-mail: `sunhwang@dju.ac.kr`

Roger Lee (Ed.): SERA 2010, SCI 296, pp. 179–195, 2010.
springerlink.com   © Springer-Verlag Berlin Heidelberg 2010

1 Introduction

The agents can be launched from a machine, navigate from web to web, collecting information or performing transactions, finally returning home with the goods or results. This scenario is especially attractive when we consider the proliferation of wireless mobile devices that is currently taking place. A user can launch an agent into the web, shutdown the device and reconnect hours later, collecting the agent with the results. Some key applications for agents in internet computing include:

- Information gathering agents, which collect information from different web sites or distributed databases, finally presenting it to its owner.
- Shopping agents, which look for the best deals for their owners, perform commercial transactions, and present the best results found, so that their owners can make a decision.
- Management agents, which carry information into selected web sites or databases and make sure that all the distributed web-infrastructure is up-to-date.
- Monitor agents, which migrate into selected web sites and monitor some information (like stock options), Warning the owner when certain events happen or even performing some actions on those events.

ad hoc network does not rely on existing infrastructure and is self-organized by the nodes which act as hosts and routers to transmit packets. By frequent changing in topology ad hoc network doesnot relay on pre-established wire network, but it requires special routing method. Routing protocols used in ad hoc networks can be divided into two categories: table-driven routing method and on-demand routing method [1].

In the table-driven routing method every node maintains information about routing of every node on the network. Nodes perform the routing based on this information. The advantage of this method is that nodes can establish the path without discovering route on need which significantly decreases performance. But periodic exchange of information between nodes for routing wastes the transmission bandwidth and nodes' energy, and creates another type of overhead. Typical table-driven routing protocols are DSDV (Destination-Sequenced Distance Vector Routing Protocol) [2], OLSR (Optimized Link State Routing Protocol). In the on-demand routing method nodes start path discovery process when data is needed to be transmitted. The advantage of this method is that nodes need not periodic information exchange for the routing which significantly decreases overhead. But process of finding routes creates transmission delay. Protocols which use on-demand routing method are AODV (Ad hoc On-demand Distance Vector) [3].

An ad hoc network is configured by many moving nodes. If an agent node that represents the network would be applicable for controlling wireless networks, it can be broadly used on organizing tree type network .

Because of overhead on table-driven routing method caused by managing routing information, on-demand routing method is studied mostly. If this overhead could be decreased somehow table-driven routing method would be acceptable to apply.

In this paper we propose a knowledge discovery agent for simple link state routing which uses bit-map information. All nodes on the network using identical bit-map information distributed by the agent that represents the network topology can decide routing path from a source to a destination at any instance. The agent collects network topology information and creates the bit-map information which shows the network topology.

The agent node can be dynamically decided using node connectivity and battery life-time information for network stability.

2 Related Works

2.1 Agent Concept Model

An agent is an atomic autonomous entity that is capable of performing some useful function. The functional capability is captured as the agent's services. A service is the knowledge level analogue of an object's operation. The quality of autonomy means that an agent's actions are not solely dictated by external events or interactions, but also by its own motivation. We capture this motivation in an attribute named purpose. The purpose will, for example, influence whether an agent agrees to a request to perform a service and also the way it provides the service. Software Agent and Human Agent are specialization of agent[4].

Figure 1 gives an informal agent-centric overview of how these concepts are inter-related. The role concept allows the part played by an agent to be separated logically from the identity of the agent itself. The distinction between role and agent is analogous to that between interface and class: a role describes the external characteristics of an agent in a particular context. An agent may be capable of playing several roles, and multiple agents may be able to play the same role. Roles can also be used as indirect references to agents. This is useful in defining re-usable patterns. Resource is used to represent non-autonomous entities such as databases or external programs used by agents. Standard object-oriented concepts are adequate for modeling resources[5].

2.2 Agents Reference Architecture on CBD

In order to construct component reference architecture, agent is classified in general agent type and mobile agent (MA) function attribute. Figure 2 is a component and meta architecture of based on all above described for Mobile Agent[6,7].

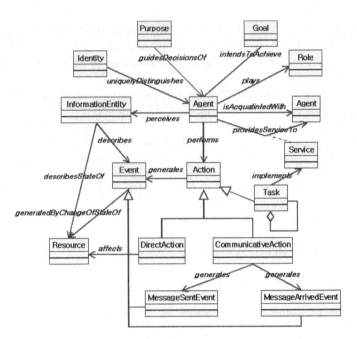

Fig. 1 Agent Concept Model

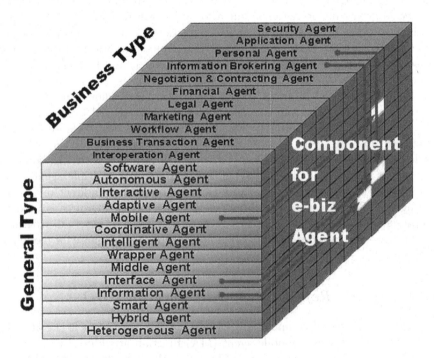

Fig. 2 Agent Reference Architecture on CBD

Reference architecture is consisted of dimension, which has 14 general types and 11 concrete business agent types with domain oriented component architecture. These two classification areas tend to be independent for each cross-referenced. Each area has its own horizontal and vertical characteristics. General agent types are corresponding to agent platform and application. It is possible to develop agent system or application by the referencing architecture. The technology of agent can be applied to business domain.

Developed component is classified by the reference architecture and is placed according to general agent type and business attribute. In case agent is applied to the agent system or business domain, system is possibly to build up by identifying component related to business domain and combining it.

In our opinion, the mobile agent paradigm provides a very good conceptual model for developing distributed internet applications. Nevertheless, there are some very important problems that must be addressed.

2.3   Creating Bit-Map and Delivery

In our proposed method the agent node gathers information for the network topology and delivers the knowledge that are represented in bit-map table to all nodes. Using simple bit-map table decreases overhead on table-driven routing. Figure 3 shows the concept of knowledge discovery agent. Using the bit-map table a source node can easily check existence of routing path to a destination node.

All nodes periodically broadcast hello message in order to find out its neighbor nodes. The agent node broadcasts query message to all neighboring nodes to collect information about network topology. The node transmitting topology query message becomes parent-node, the receiver becomes child-nodes.

The node which has received topology query message delivers it to its child-nodes. If it has no child-node, i.e. it is a leaf node, than it sends topology reply message to parent-node which includes the followings: node ID, connection and battery life-time information. By repeating this sequence the agent node collects all nodes connection information of the network. After finishing the query process the agent node creates topology knowledge using all this information as a bit-map table.

On the network shown in Figure 4, the agent node is node 1 that collects connection information on the network and creates the topology knowledge as the bit-map table.

The network topology represented as a bit-map table is shown in Table 1 where 1 and 0 stand for one-hop connection and no one-hop connection respectively.

If a new node comes in or leaves the network or any alteration occurs on the network topology, all neighbour nodes that have sensed the change send topology reply message to agent node. The agent node has to maintain the

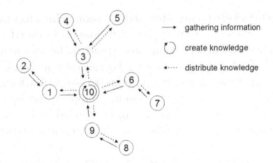

Fig. 3 Knowledge Discovery Agent

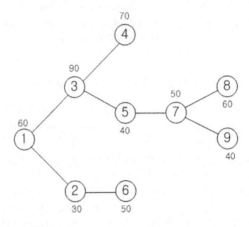

Fig. 4 A network topology

bit-map table and transmit the new bit-map table information to all other nodes.

The agent node elects substitute agent nodes using all nodes connectivity and battery-life time information. In case the agent node disappears from the network, a new agent node has to be selected [8]. Then it sends bit-map and the new agent node information via topology advertisement message to all nodes. In this way every node on the network will know about existence of the agent and all connections on the network. An arbitrary node can be the agent node when the network is initially created. In this initial case, selecting process for the proper agent node is main duty.

2.4 Agent Selection

The agent selection algorithm is based on the leader election algorithm. We propose a list of candidate agents instead of just one agent to be additionally maintained in every node. Each node contains an agent list of five nodes (in descending order), where the first node is considered as the active agent of the network. When the agent node disappears, the highest-valued?node from the collected agent is selected as the substitute node. [9]. Network partition and merging can also be managed efficiently by our proposed method.

3 Bit-Map Based Routing Protocol

All nodes on the network periodically send Hello message in order to confirm existence of neighbor nodes.

After receiving hello message from neighbor nodes, a node builds up topology information table. Topology information table consists of information of neighbor nodes and node's own information.

For organizing network topology, the agent node broadcasts topology query message to all neighbour nodes. Figure 5 shows the format of query message. By this way it builds self-centred topology tree.

Table 1 Bit-map table for network topology shown on Figure 1

Node ID	1	2	3	4	5	6	7	8	9
1	0	1	1	0	0	0	0	0	0
2	1	0	0	0	0	1	0	0	0
3	1	0	0	1	1	0	0	0	0
4	0	0	1	0	0	0	0	0	0
5	0	0	1	0	0	0	1	0	0
6	0	1	0	0	0	0	0	0	0
7	0	0	0	0	1	0	0	1	1
8	0	0	0	0	0	0	1	0	0
9	0	0	0	0	0	0	1	0	0

The node which has received topology query message assigns the node which has sent the message as parent-node, and all other neighbor nodes are assigned as child-node.

A node after receiving topology query message, it sends topology reply message to parent-node after gathering sub-tree information. The message contains nodes ID, battery life-time and connectivity information.

A parent node after receiving all topology reply messages from its child-nodes records all information on topology information table and transmits the information to its parent-node. Figure 6 shows the format of topology reply message.

The parent-node sends topology reply message to its parent-node only after it receives topology reply message from all its child-nodes. Using this method, the agent node can accumulate topology information for all nodes on the network.

Nodes 4, 6, 8 and 9 on Figure 2 have no child-nodes, thus they will promptly send topology reply message to their parent-nodes. Topology information table of node 5 after receiving topology reply message is shown on Figure 7.

After receiving topology information the agent node builds up bit-map table. Based on the bit-map table, the agent node can be changed by the following rule:

a. Leaf-node can not be the agent node.
b. The agent node should have longest battery life-time.
c. The agent node should have as many as possible neighbor nodes.

By applying above rules node 3 is the best for the agent node. Nodes 1 and 2 can be candidates. In this case, the agent node delivers collected bit-map table which represents network topology to the new agent node.

The agent node sends the bit-map information to all nodes on the network by topology advertisement message. Figure 8 shows topology advertisement message format.

Nodes that have received topology message checks whether there exists difference between its own previous topology information. Then, the node updates the topology table.

The agent node periodically broadcasts topology advertisement message to all other nodes which have to know the recent information about network topology and the agent node.

If a node needs to transmit data to another node then it can check a routing path to the destination node by using bit-map table and can quickly establish the connection. The bit-map table represent a graph which shows network topology. The pseudo algorithm of the process is illustrated on Figure 9.

Type	T query ID	Reserved
Agent IP address		
Agent sequence Number		

Fig. 5 Topology query message format

Type	T_reply ID	Reserved
Node's own IP address		
Battery	Connectivity	Agent
Neighbor Node's IP address		
Battery	Connectivity	Agent
.		
.		
.		
Agent IP address		

Fig. 6 Topology reply message format

Node ID	Neighbor Node	Battery	Connection information	Agent
5	3(1), 7(0)	40	2	
7	5(1), 8(0), 9(0)	50	3	
8	7(1)	60	1	
9	7(1)	40	1	

Fig. 7 Topology information of node 5

Type	T_adv ID	Reserved
Agent node ID		
1st candidate-agent node ID		
2nd candidate-agent node ID		
3rd candidate-agent node ID		
4th candidate-agent node ID		
5th candidate-agent node ID		
Bit-map table		

Fig. 8 Topology message format

```
1.   Do  "AND"  operation  between  source  node  and
     destination node.
     A.  If there is a path to the destination node, then
         go to 3
     B.  If there is no path, then go to step 2
2.   Find neighbor node.
     A.  Select a neighbor node, do "AND" operation with
         destination.
         i.     Finding a path to destination, then go to 3.
         ii.    If there is no path, go to 2-A.
     B.  When all neighbors are checked, then select a
         neighbor and do 2-A recursively.
3.   The searched nodes are used for path, and finish.
```

Fig. 9 Algorithm of path discovery

At first source node performs AND operation between itself and the destination node. If the source node finds connection between itself and the destination node, then the connection is established immediately.

If no connection is found between the source and the destination node, the source node searches for neighbor node. The neighbor nodes become intermediate nodes as the relay to the destination. Then AND operation is repeated between neighbor nodes and the destination node. If no connection is found between neighbor nodes and the destination node, the neighbor node searches for its neighbor nodes. And then AND operation is performed to the destination node. This process is continued until a connection is found to the destination.

The method explained above is used for finding a path from the source to the destination node. Multipaths can also be established between the source and the destination nodes. The other paths could be assigned as reference paths.

4 Verification of Route Discovery

For verification of the proposed agent method, an example ad hoc network is shown at Figure 10. The created bit-map is shown in Table 2. The agent can be decided by using the method of 2.2.

```
1.  Do "AND" operation between source node and
    destination node.
    A. If there is a path to the destination node, then
       go to 3
    B. If there is no path, then go to step 2
2.  Find neighbor node.
    A. Select a neighbor node, do "AND" operation with
       destination.
    i.    Finding a path to destination, then go to 3.
    ii.   If there is no path, go to 2-A.
    B. When all neighbors are checked, then select a
       neighbor and do 2-A recursively.
3.  The searched nodes are used for path, and finish.
```

Fig. 10 An experimental network topology

Table 2 Bit-map table for network shown on Figure 10

Node ID	1	2	3	4	5	6	7	8	9			
1	0	1	1	1	1	0	0	0	0	0	0	0
2	1	0	1	0	0	1	0	0	0	0	0	0
3	1	1	0	1	0	0	1	0	0	0	0	0
4	1	0	1	0	0	0	1	0	0	0	0	0
5	1	0	0	0	0	0	0	1	0	0	0	0
6	0	1	0	0	0	0	0	0	0	1	0	0
7	0	0	1	1	0	0	0	0	1	0	0	1
8	0	0	0	0	1	0	0	0	0	0	1	0
9	0	0	0	0	0	0	1	0	0	0	0	1
10	0	0	0	0	0	1	0	0	0	0	0	1
11	0	0	0	0	0	0	0	1	0	0	0	1
12	0	0	0	0	0	0	1	0	1	1	1	0

```
NODE 1
Input SRC & DST: 1 2    1 -> 2

Input SRC & DST: 1 3    1 -> 3

Input SRC & DST: 1 4    1 -> 4

Input SRC & DST: 1 5    1 -> 5

Input SRC & DST: 1 6    1 -> 2 -> 6

Input SRC & DST: 1 7    1 -> 3 -> 7,    1 -> 4 -> 7

Input SRC & DST: 1 8    1 -> 5 -> 8

Input SRC & DST: 1 9    1 -> 3 -> 7 -> 9,    1 -> 4 -> 7 -> 9

Input SRC & DST: 1 10   1 -> 2 -> 6 -> 10

Input SRC & DST: 1 11   1 -> 5 -> 8 -> 11

Input SRC & DST: 1 12   1 -> 3 -> 7 -> 12
```

Fig. 11 Route searching procedure

Node 1 is set as a source node. For finding all destinations bit-map operation is performed by the algorithm of Figure 9. The operation results are shown on Figure 11. It shows that paths are established for all destinations. It also shows multiple paths with different number of hops for the same destination.

For data transmission a path with minimum number of hops will be chosen. In practice, when there are multiple paths with same number of hops, one of them will be chosen for the primary path and others will be reserved as substitutions. We have experimented on various topologies. The probability of finding destinations is 100.

5 Verification of Route Discovery

The effectiveness of proposed method is compared with AODV.

5.1 Agent Concept Model

- Number of nodes: 40,60, 80 respectively;
- Testing area: 1000m x 1000m;
- Mobile node speed: varies between 0 to 2 m/s;
- Mobility model: random way point model (when the node reaches its destination, it pauses for several seconds, e.g., 1s, then randomly chooses another destination point within the field, with a randomly selected constant velocity);
- Traffic load: UDP, CBR traffic generator;
- Radio transmission range: 250 m; and
- MAC layer: IEEE 802.11.

Each simulation is run for 500 seconds and each case has 10 different topologies.

5.2 Simulation Results

Figure 12 shows packet deliver ratio of AODV and BITMAP. Difference between these protocols is not big.

Figure 13 shows the control packet overhead required for transferring the routing packets. AODV has less control packet overhead while the number of nodes is few. But, when the number of nodes increases BITMAP has lower control overhead.

Figure 14 shows the average energy remained of each protocol. We have to mention that it is a mean value of energy remained each node at the end of simulation. Remained energy in BITMAP is higher than AODV when the number of nodes increases.

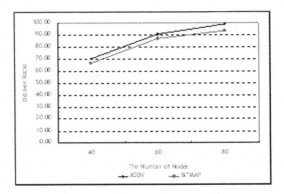

Fig. 12 Packet Deliver Ratio (no mobility, 3 connections)

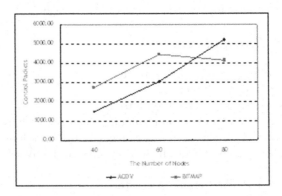

Fig. 13 Control Packet Overhead (no mobility, 3 connections)

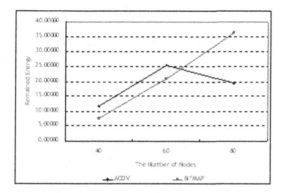

Fig. 14 Average Remained Energy (no mobility, 3 connections)

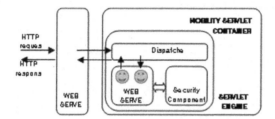

Fig. 15 The mobility KDA container

5.3 Development of Knowledge Discovery Agent

The implementation of the KDA(Knowledge Discovery Agent) container follows the same base guidelines of the Mobility wrapper, but with same important changes.

First, the web server may be decoupled from the KDA engine, and from the KDA itself. In this case, the function of the web server is to provide a mapping between URIs and the resources, forwarding the requests to the appropriate KDA. Each request that corresponds to an interaction with an agent, is forward to the Mobility KDA Container, which then passes it to the appropriate agent. Secondly, the Security Component of our framework is instantiated and running, providing security features for the running agents and for the host. Figure 15 shows the approach of the mobility KDA container. It should be noted that it is not necessary to decouple the web server from the KDA engine idea was used in our implementation. Figure 16 and 17 show the modeling process using UML to implement the KDA.

As shown in the figure, when an agent arrives at a web server, it may not only query the local web server, but it can also ask for a service instance that allows it to behave as a KDA, and publish information. This also enables the agents to perform maintenance tasks on it from the inside. In fact, one of our first prototypes was a simple management application that used mobile

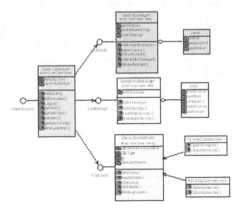

Fig. 16 The mobility KDA Use Case

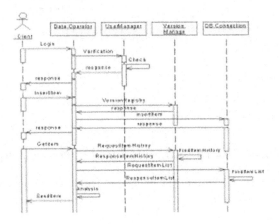

Fig. 17 The mobility KDA State diagram

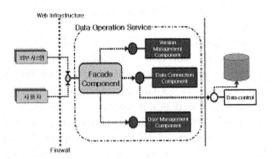

Fig. 18 The mobility KDA using CBD

agents to collect information from the web servers, and performed simple administration operations on them. When an agent requires an object that allows it to publish information, the object that is passed actually requires that the agent to implement the KDA interface. Figure 18 show the execution environment of KDA that made to an agent contains the full information about the request. This includes not only the IP of the client, MIME-types accepted but also session information, This is important since allows the agents to distinguish between different clients, and act accordingly. Also, when the agents request the service instance, they can require or deny that they are listed on line.

The figure 19 shows the sample execution of the KDA on web.

Fig. 19 Sample of Execution mobility KDA

6 Conclusions

Designing, developing and maintaining components for reuse is a very complex process, which require not only for the component functionality and flexibility, but also for matured development organization. We have proposed a knowledge discovery agent for gathering network information to increase performance on ad hoc networks. The concept of proposed agent and an algorithm for knowledge of network topology are provided. The bit-map knowledge from collected information that includes routing information of the network is created and distributed by the agent. The bit-map knowledge is the main of routing at each node. Computer simulation on various network topologies shows that suggested knowledge discovery agent method for making bit-map method help sources find destinations with 100% probability.

The bit-map table could be also used to search multi-paths. Our KDA also provides a very strong security model, with authentication and authorization mechanisms that control incoming agents, and cryptographic primitives that are useful to protect the integrity and confidentiality of the agents. This is essential if an infrastructure is going to be deployed on an open environment.

Finally, our framework also allows the user distrust on mobile agents to be addressed. With our framework the user does not even has to be aware that he is using mobile agent technology. He only should be aware of the positive results of using this technology.

Further researches should aim to design concrete protocol and compare its performance with the existing ad hoc network routing protocols. Effective shortest path finding method has also to be studied.

Acknowledgements. This work was supported by the Korea National Research Foundation (NRF) granted funded by the Korea Government (MEST No. R- 2009-0083879).

References

1. Agent Platform Special Interest Group, Agent Technology green paper, OMG Document agent/00-0-01 Version 1.0 (2000),
   http://www.objs.com/agent/index.html
2. Johansen, D., Lauvset, K.J., Marzullo, K.: An Extensible Software Architecture for Mobile Components. In: Proceeding of the Ninth Annual IEEE International Conference and Workshop on the Engineering of Computer-Based Systems, pp. 231–237 (2002)
3. Tan, T.-H., Liu, T.-Y.: The MObile-Based Interactive Learning Environment (MOBILE) and A Case Study for Assisting Elementary School English Learning. In: Proceedings of IEEE International Conference on Advanced Learning Technologies (ICALT 2004), August 2004, pp. 530–534 (2004)
4. Ogier, R.G., Lewis, M.G., Templin, F.L.: Topology Broadcast Based on Reverse-Path Forwarding. draft-ietf-manet-tbrpf-08.txt (April 2003)
5. Johnson, D., Maltz, D., Hu, Y.-C.: The Dynamic Source Routing Protocol for Mobile Ad Hoc Networks (DSR). IETF Mobile Ad Hoc Networks Working Group, Internet Draft, April 15 (2003) (work in progress)
6. Nasipuri, A., Castaneda, R., Dasl, S.: Performance of Multipath Routing for On-demand Protocols in Mobile Ad hoc Networks. Kluwer Academic Publishers Mobile Networks and Applications (2001)
7. Odell, J., Van Dyke Parunak, H., Bauer, B.: Extending UML for Agents. In: Proceeding of Agent-Oriented Information Systems Workshop at the 17th International Conference on Artificial Intelligence, vol. 11(3), pp. 303–328 (2001)
8. Chainbi, W.: Using the Object paradigm to deal with the agent paradigm: capabilities and limits. In: Proceeding of 2001 ACM Symposium on Applied Computing, March 2001, pp. 585–589 (2001)
9. Depke, R., Heckel, R., Kuster, J.M.: Integrating Visual Modeling of Agent-Based and Object-Oriented Systems. In: Proceeding of Forth International Conference on Autonomous Agents 2000, June 2000, pp. 82–83 (2000)

10. Bauer: UML Class Diagrams and Agent-Based Systems. In: Proceeding of Fifth International Conference on Autonomous Agents 2001, May 2001, pp. 104–105 (2001)

11. Shao, W., Tasi, W.-T., Rayadurgam, S., Lai, R.: An Agent Architecture for Supporting Individualized Services in Internet Applications. In: Proceeding of Tenth IEEE International Conference on Tools with Artificial Intelligence, November 1998, vol. 10(12), pp. 140–147 (1998)

Aspect-Oriented Modeling for Representing and Integrating Security Concerns in UML

D. Mouheb, C. Talhi, M. Nouh, V. Lima, M. Debbabi,
L. Wang, and M. Pourzandi

Abstract. Security is a challenging task in software engineering. Enforcing security policies should be taken care of during the early phases of the software development process to more efficiently integrate security into software. Since security is a crosscutting concern that pervades the entire software, integrating security at the software design level may result in the scattering and tangling of security features throughout the entire design. To address this issue, we present in this paper an aspect-oriented modeling approach for specifying and integrating security concerns into UML design models. In the proposed approach, security experts specify high-level and generic security solutions that can be later instantiated by developers, then automatically woven into UML design. Finally, we describe our prototype implemented as a plug-in in a commercial software development environment.

1 Introduction

With pervasiveness of computer systems in all aspects of human activities, software complexity is increasing drastically, resulting in the interest on generating robust code from high level design languages like UML. In this context, we propose an approach for representing and integrating security aspects in UML. Our work is part of a research project (MOBS2) on the model-based engineering of secure software and systems. This project aims at providing an end-to-end framework for secure software development. This means a framework that starts from the specification of

D. Mouheb, C. Talhi, M. Nouh, V. Lima, M. Debbabi, and L. Wang
Computer Security Laboratory, Concordia University, Montreal, Canada
e-mail: {d_mouheb,talhi,m_nouh,v_nune}@ciise.concordia.ca,
{debbabi,wang}@ciise.concordia.ca

M. Pourzandi
Software Research, Ericsson Canada Inc., Montreal, Canada
e-mail: makan.pourzandi@ericsson.com

Roger Lee (Ed.): SERA 2010, SCI 296, pp. 197–213, 2010.

the needed security requirements on UML models, verification and validation of the models against the specified requirements, enforcing those requirements on UML models, and ends with secure code generation. In this paper, we focus on enforcing the needed security requirements using Aspect-Oriented Modeling (AOM) [1].

In fact, AOM has become the center of many recent research activities [23,7,16, 9,10,22,4]. AOM allows software developers to conceptualize and express concerns in the form of aspects at the modeling stage, and integrate them into their UML diagrams using UML composition techniques. The usefulness of aspect-oriented techniques for enforcing security requirements in software systems has been already demonstrated in the literature [20,2,21]. Though, in spite of the increasing interest, to date, there is no standard language to support AOM, nor a standard mechanism for weaving aspects into the base models.

In this paper, we provide a new approach for specifying security aspects in UML. In addition, our approach allows systematically and automatically weaving security aspects into UML design models and therefore enabling secure code generation. By systematically, we mean that our approach supports a complete coverage of the main UML diagrams that are used in software design (class diagrams, state machine diagrams, sequence diagrams, and activity diagrams), and by automatically, we mean that our approach creates the composed new UML models from aspects and original UML models based on formal rules defined in our approach with less intervention from the developer. In the proposed approach, the security expert specifies the needed security solutions as application-independent aspects. In addition, he/she specifies how those aspects should be integrated into the design. The developer then specializes the application-independent aspects to his/her design. Finally, our framework injects the application-dependent aspects at the appropriate locations in the design models. The main contributions of our work are the following:

- We devise a UML profile that assists security experts in specifying security solutions as aspects.
- We define a pointcut language to designate the locations where those aspects should be injected into base models.
- We introduce a new weaving interface for developers to specialize the generic aspects provided by the security expert in order to adapt them to their application.
- We design and implement a weaving mechanism based on model-to-model transformation to inject automatically, i.e., without user intervention, aspects into UML models.

This paper extends the work done in [14] by proposing a more expressive and generic AOM approach for representing aspects, their corresponding adaptations and pointcuts, and new weaving mechanisms. Furthermore, we discuss our prototype implemented as a plug-in to the IBM Rational Software Architect tool [12].

The remainder of this paper is organized as follows. Section 2 summarizes our approach for specifying and weaving aspects into UML design models. Section 3 introduces an example that will be used to illustrate the usability of the approach. Afterwards, Section 4 presents our UML profile for AOM. The weaving approach is presented in Section 5. Section 6 shows the feasibility of the approach by applying it to the example of Section 3. Section 7 gives an overview of the related work. Finally, we conclude the paper and present our future work in Section 8.

2   Enforcing Security Requirements in the MOBS2 Framework

Security as a non-functional requirement can be modeled as an aspect. The process of applying the aspect to the base model is commonly called "Weaving". Figure 1 presents a high level overview of the proposed approach for specifying and weaving security aspects in the MOBS2 framework.

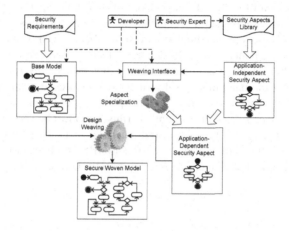

Fig. 1 Proposed Approach: Specification and Weaving of UML Security Aspects.

In the MOBS2 framework, the security expert has the responsibility of designing the application-independent aspects. By analogy, these aspects are generic templates representing the security features independently of the application specificities. In order to assist security experts in designing the security aspects, a UML profile was developed in MOBS2 such that aspects can be specified by attaching stereotypes, parameterized by tagged values, to UML elements (See Section 4). Additionally, we developed a high level language to present the pointcuts that specify the locations in the base model where the aspects should be applied (See Subsection 4.3).

The developer in turn has the responsibility to specialize the application-independent aspects provided by the security expert according to the application-specific security requirements and needs. The developer must specify where to integrate security mechanisms in the base model through MOBS2 weaving interface (See Subsection 5.1). Based on the pointcuts specified in the aspect by the security expert and specialized by the developer, the MOBS2 framework identifies and selects, without any developer interaction, the join points from the base model where the aspect adaptations should be performed (See Subsection 5.2).

At the end, the MOBS2 framework automatically weaves the aspects into the base model. To provide a portable solution, we adopted a model-to-model transformation language; the QVT language [5]. For each adaptation, a set of QVT rules are generated by MOBS2 framework. The order of applying these rules to the base model satisfies the semantics of the original aspect adaptations (See Subsection 5.3).

3   Use Case Study: RBAC Support

This section presents a service provider application that will be used to illustrate our approach. The use case consists of enforcing access control to different application resources based on Role-Based Access Control (RBAC) [18]. The class diagram of the service provider application is depicted in Figure 2. The class *Client* represents the application's users (e.g., administrator, subscribers). Each type of user has specific privileges. A client accesses the database of subscribers (*ResourceDB*) through an interface *Provision* that is implemented by the classes *SubscriberManager* and *ServiceManager* for manipulating subscribers and services respectively.

Figure 3 represents a sequence diagram specifying the behavior of the method *SubscriberManager.delete()*. The client's permissions must be verified before deleting a subscriber (i.e., only the administrator can delete a subscriber). In the following sections, we show how the designer uses MOBS2 framework to weave RBAC aspect into the base model to check user permissions before deleting a subscriber.

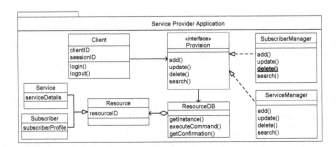

Fig. 2 Class Diagram for a Service Provider Application.

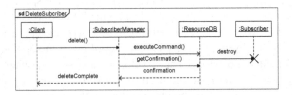

Fig. 3 Sequence Diagram Specifying the Behavior of the Method *SubscriberManager.delete()*.

4 A UML Profile for Aspect-Oriented Modeling

This section presents our UML profile used for representing aspects in UML design. An aspect in MOBS2 contains a set of adaptations and pointcuts. An adaptation specifies the modification that an aspect performs on the base model. A pointcut specifies the locations in the base model (join points in AOP) where an adaptation should be performed. The elements of this profile will be used by security experts to specify security solutions for well-known security problems. However, the profile is generic enough to be used for specifying non-security aspects. In our profile, an aspect is represented as a stereotyped package. For example, Figure 5 shows a partial specification of the RBAC aspect. It is modeled as a package stereotyped ≪*aspect*≫. In the following subsections, we show how adaptations and pointcuts can be specified using our AOM profile.

4.1 Aspect Adaptations

As mentioned earlier, an adaptation specifies the modification that an aspect performs on the base model. We classify adaptations according to the covered diagrams and the adaptation rules. We focus on the diagrams that are the most used by developers: class diagrams, sequence diagrams, state machine diagrams, and activity diagrams. Figure 4 presents our specification of adaptations. We define two types of adaptations: structural and behavioral adaptations.

4.1.1 Structural Adaptations

Structural adaptations specify the modifications that affect structural diagrams. We focus on class diagrams since they are the most used structural diagrams in the software design. A class diagram adaptation is similar to an introduction in AOP languages (e.g., AspectJ). A structural adaptation is modeled as an abstract meta-element named *StructuralAdaptation*. It is specialized by the meta-element *ClassAdaptation* used to specify class diagram adaptations. For example, *RoleAddition* in Figure 5 is a class adaptation used for the integration of a class *Role* into the class diagram of the service provider application (Figure 2).

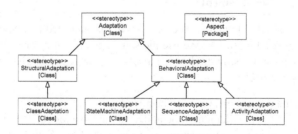

Fig. 4 Meta-language for Specifying Aspects and their Adaptations.

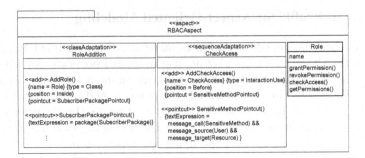

Fig. 5 Partial View of the RBAC Aspect.

4.1.2 Behavioral Adaptations

Behavioral adaptations specify the modifications that affect behavioral diagrams. In our approach, we support the behavioral diagrams that are the most used for the specification of a system behavior, mainly, state machine, sequence, and activity diagrams. A behavioral adaptation is similar to an advice in AOP languages (e.g., AspectJ). A behavioral adaptation is modeled as an abstract meta-element named *BehavioralAdaptation*. We specialized the meta-element *BehavioralAdaptation* by three meta-elements: *StateMachineAdaptation*, *SequenceAdaptation*, and *ActivityAdaptation* that are used to specify adaptations for state machine, sequence, and activity diagrams respectively. For example, *CheckAccess* in Figure 5 is a sequence adaptation defining the adaptation rules required to inject the behavior needed to check user permissions before any call to a sensitive method.

4.2 Aspect Adaptation Rules

An adaptation rule specifies the effect that an aspect performs on the base model elements. We support two types of adaptation rules: *adding* a new element to the base model and *removing* an existing element from the base model. Figure 6 depicts our specified meta-model for adaptation rules.

4.2.1 Adding a New Element

The addition of a new element to the base model is modeled as a special kind of operation stereotyped ≪*Add*≫. We use the same specification for adding any kind of UML element, either structural or behavioral. Three tagged values are attached to the stereotype ≪*Add*≫:

- *Name:* The name of the element to be added.
- *Type:* The type of the element to be added to the base model. The values of this tag are provided in the enumerations *ClassElementType*, *StateMachineElementType*, *SequenceElementType*, and *ActivityElementType*.

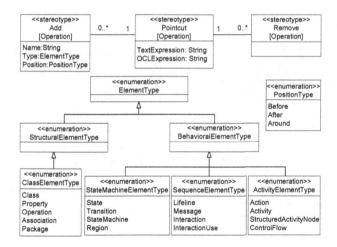

Fig. 6 Meta-language for Specifying Adaptation Rules.

- *Position:* The position where the new element needs to be added. The values of this tag are given in the enumeration *PositionType*. This tag is optional for some elements (e.g., a class, an operation) since these kinds of elements are always added inside a join point.

The location where the new element should be added is specified by the meta-element *Pointcut* (Section 4.3). For example, the stereotyped operation *AddRole()* in Figure 5 is an adaptation rule specifying the addition of a new class *Role* to the package *SubscriberPackage* matched by the pointcut *SubsriberPackagePointcut*.

4.2.2   Removing an Element

The deletion of an existing element from the base model is modeled as a special kind of operation stereotyped ≪*Remove*≫. The elements that should be removed are given by a pointcut specified by the meta-element *Pointcut* (Subsection 4.3). The same specification is used for removing any kind of UML element, either structural or behavioral. No tagged value is required for the specification of a *Remove* adaptation rule; the pointcut is enough to select the elements that should be removed.

4.3   Pointcuts

A pointcut is an expression that allows the selection of a set of locations in the base model (join points in AOP jargon) where adaptations should be performed. Since the targeted join points are UML elements, pointcuts should be defined based on designators that are specific to the UML language. To this end, we defined in our approach a pointcut language that provides UML-specific pointcut designators needed to select UML join points. The proposed pointcut language covers all the

Table 1 Class Diagram Pointcuts.

Join Point	Pointcut Designator	Description
Class	Class(NamePattern)	Selects a class based on its name.
	Inside_Package(PackagePointcut)	Selects a class that belongs to a specific package.
	Contains_Attribute(AttributePointcut)	Selects a class that contains a specific attribute.
	Contains_Operation(OperationPointcut)	Selects a class that contains a specific operation.
	Associated_With(ClassPointcut)	Selects a class that is associated with a specific class.
Attribute	Attribute(NamePattern)	Selects an attribute based on its name.
	Inside_Class(ClassPointcut)	Selects an attribute that belongs to a specific class.
	Of_Type(TypePattern)	Selects an attribute that is of certain type.
	Of_Visibility(VisibilityKind)	Selects an attribute that is of certain visibility.
Operation	Operation(NamePattern)	Selects an operation based on its name.
	Inside_Class(ClassPointcut)	Selects an operation that belongs to a specific class.
	Args(TypePattern1, TypePattern2, ...)	Selects an operation based on the type of its arguments.
	Of_Visibility(VisibilityKind)	Selects an operation that is of certain visibility.
Association	Association(NamePattern)	Selects an association based on its name.
	Between(ClassPointcut, ClassPointcut)	Selects an association that is between certain classes.
	Member_Ends(AttributePointcut, AttributePointcut)	Selects an association based on its member ends.
	Aggregation_Kind(AggregationKind)	Selects an association based on its aggregation kind.
Package	Package(NamePattern)	Selects a package based on its name.
	Inside_Package(PackagePointcut)	Selects a package that belongs to a specific package.
	Contains_Class(ClassPointcut)	Selects a package that contains a specific class.

Table 2 Sequence Diagram Pointcuts.

Join Point	Pointcut Designator	Description
Interaction	Interaction(NamePattern)	Selects an interaction based on its name.
	Contains_Message(MessagePointcut)	Selects an interaction that contains a specific message.
	Contains_Lifeline(LifelinePointcut)	Selects an interaction that contains a specific lifeline.
	Specifies_Operation(OperationPointcut)	Selects an interaction that specifies the behavior of a specific operation.
Message	Message_Call(NamePattern)	Selects a message call, either synchronous or asynchronous, based on its name.
	Message_Syn_Call(NamePattern)	Selects a message that specifies a synchronous call.
	Message_Asyn_Call(NamePattern)	Selects a message that specifies an asynchronous call.
	Replay_Message(NamePattern)	Selects a replay message based on its name.
	Create_Message(NamePattern)	Selects a message that creates an object.
	Destroy_Message(NamePattern)	Selects a message that destroys an object.
	Message_Source(TypePattern)	Selects a message whose source is of certain type.
	Message_Target(TypePattern)	Selects a message whose target is of certain type.
	Inside_Interaction(InteractionPointcut)	Selects a message that belongs to a specific interaction.
Lifeline	Lifeline(NamePattern)	Selects a lifeline based on its name.
	Inside_Interaction(InteractionPointcut)	Selects a lifeline that belongs to a specific interaction.
	Covered_By_Fragment(NamePattern)	Selects a lifeline that is covered by a specific interaction fragment.
	Contains_Execution(NamePattern)	Selects a lifeline that contains a specific execution specification.

kinds of join points where the adaptations supported by our approach are performed. In the following tables (Table 1 through Table 4), we present the primitive pointcut designators for the main UML diagrams that are supported by our approach (i.e., class diagrams, sequence diagrams, activity diagrams, and state machine diagrams).

Those primitive pointcut designators can be composed with logical operators (AND, OR, and NOT) to build other pointcuts. For example, the pointcut *SensitiveMethodPointcut* in Figure 5 is a conjunction of three pointcuts: (1) *Message_Call(SensitiveMethod)* selects any message in the base model that calls *SensitiveMethod()*, (2) *Message_Source(User)* selects any message whose source is of type *User*, and (3) *Message_Target(Resource)* selects any message whose target is of type *Resource*. The conjunction of these three pointcuts allows the selection of all message calls to *SensitiveMethod()* from a *User* instance to a *Resource* instance.

The proposed pointcut language is expressive enough to designate the main UML elements that are used in a software design. For example, the pointcut language allows to designate a package, a class, an operation, an attribute, or an association in a class diagram, a message, a lifeline, or an interaction in a sequence diagram, etc.

Table 3 Activity Diagram Pointcuts.

Join Point	Pointcut Designator	Description
Activity	Activity(NamePattern)	Selects an activity based on its name.
	Contains_Action(ActionPointcut)	Selects an activity that contains a specific action.
	Contains_Edge(EdgePointcut)	Selects an activity that contains a specific activity edge.
	Specifies_Operation(OperationPointcut)	Selects an activity that specifies the behavior of a specific operation.
Action	Action(NamePattern)	Selects an action based on its name.
	Call_Operation_Action(NamePattern)	Selects an action that performs an operation call.
	Call_Behavior_Action(NamePattern)	Selects an action that performs a behavior call.
	Create_Action(NamePattern)	Selects an action that creates an object.
	Destroy_Action(NamePattern)	Selects an action that destroys an object.
	Read_Action(NamePattern)	Selects an action that reads the value(s) of a structural feature.
	Write_Action(NamePattern)	Selects an action that updates the value(s) of a structural feature.
	Inside_Activity(ActivityPointcut)	Selects an action that belongs to a specific activity.
	Input(TypePattern, ...)	Selects an action based on the type of its input pins.
	Output(TypePattern, ...)	Selects an action based on the type of its output pins.
Activity Edge	Edge(NamePattern)	Selects an edge based on its name.
	Inside_Activity(ActivityPointcut)	Selects an edge that belongs to a specific activity.
	Source_Action(ActionPointcut)	Selects an edge that has a specific source action.
	Target_Action(ActionPointcut)	Selects an edge that has a specific target action.

Table 4 State Machine Diagram Pointcuts.

Join Point	Pointcut Designator	Description
State Machine	State_Machine(NamePattern)	Selects a state machine based on its name.
	Contains_Region(RegionPointcut)	Selects a state machine that contains a specific region.
	Contains_State(StatePointcut)	Selects a state machine that contains a specific state.
	Contains_Transition(TransitionPointcut)	Selects a state machine that contains a specific transition.
	Specifies_Class(ClassPointcut)	Selects a state machine that specifies a specific class.
Region	Region(NamePattern)	Selects a region based on its name.
	Inside_State_Machine(StateMachinePointcut)	Selects a region that belongs to a specific state machine.
	Inside_State(StatePointcut)	Selects a region that belongs to a specific state.
	Contains_State(StatePointcut)	Selects a region that contains a specific state.
	Contains_Transition(TransitionPointcut)	Selects a region that contains a specific transition.
State	State(NamePattern)	Selects a state based on its name.
	Inside_Region(RegionPointcut)	Selects a state that belongs to a specific region.
	Inside_State(StatePointcut)	Selects a state that belongs to a specific state.
	Inside_State_Machine(StateMachinePointcut)	Selects a state that belongs to a specific state machine.
	Incoming(TransitionPointcut)	Selects a state that has a specific incoming transition.
	Outgoing(TransitionPointcut)	Selects a state that has a specific outgoing transition.
	Contains_State(StatePointcut)	Selects a state that contains a specific state.
	Contains_Transition(TransitionPointcut)	Selects a state that contains a specific transition.
Transition	Transition(NamePattern)	Selects a transition based on its name.
	Inside_Region(RegionPointcut)	Selects a transition that belongs to a specific region.
	Inside_State(StatePointcut)	Selects a transition that belongs to a specific state.
	Inside_State_Machine(StateMachinePointcut)	Selects a transition that belongs to a specific state machine.
	Source_State(StatePointcut)	Selects a transition that has a specific source state.
	Target_State(StatePointcut)	Selects a transition that has a specific target state.

In addition, our proposed pointcut language provides high-level and user-friendly primitives that can be used by the security expert to designate UML elements.

5 Weaving UML Aspects Using M2M Transformation

Model Transformation is a concept defined by the Object Management Group (OMG) as: "the process of converting one model to another model of the same system" [13]. This process takes as input one or more models that conform to a specific meta-model and produces as output one or more models that conform to a given meta-model. Model transformation can be instrumented to achieve the goals of model weaving. Figure 8 presents an overview of the proposed weaving approach.

This high level overview presents the steps and the technologies used to weave security aspects into UML design models. Having an aspect designed using the AOM profile, the textual pointcuts will first be translated into equivalent OCL pointcuts.

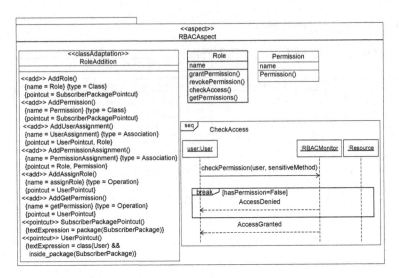

Fig. 7 Specification of the RBAC Aspect.

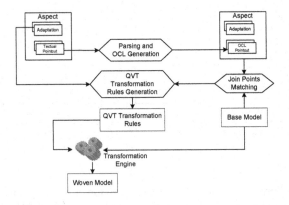

Fig. 8 Overview of the Proposed Weaving Approach.

Based on the generated OCL pointcuts, the join points will be identified through the join points matching process (See Section 5.2). The aspect adaptations will be translated into equivalent QVT transformation rules by the QVT Transformation Rules Generation. These rules in turn will be executed by the Transformation Engine to generate a secure woven model. In the following subsections, we will explain each process of the proposed weaving approach.

5.1 Aspects Specialization through a Weaving Interface

During this step, the developer specializes the generic aspects by choosing the elements of his/her model that are targeted by the security solutions. The pointcuts specified by security experts are chosen to match specific points of the design where security methods should be added. Since security solutions are provided as a library of aspects, pointcuts are specified as generic patterns that should match all possible join points that can be targeted by the security solutions. To specialize the aspects, we provide a weaving interface that hides the complexity of the security solutions and only exposes the generic pointcuts to the developers. From this weaving interface and based on his/her understanding of the application, the developer has the possibility of mapping each generic element of the aspect to its corresponding element(s) in the base model. After mapping all the generic elements, the application-dependent aspect will be automatically generated.

5.2 Join Points Identification

During this step, the actual join points where the aspect adaptations should be performed are selected from the base model. To select the targeted join points, the textual pointcuts presented in Subsection 4.3 need to be translated to a language that can navigate the base model elements. In our approach, we use the standard OCL as a query language. We translate textual pointcuts to OCL constraints which serve as predicates to select the considered join points. This translation is done using CUP Parser Generator for Java. This tool takes as input: (1) the grammar of the pointcut language along with the actions required to translate each primitive pointcut to its corresponding OCL pointcut, and (2) a scanner used to break the pointcut expression into meaningful tokens, and provides as output a parser that is capable of parsing and translating any textual pointcut expression to its equivalent OCL expression. Once the OCL expression is generated, it will be evaluated on the base model to select the targeted join points. For example, the pointcut *UserPointcut* presented in Figure 7 with the textual expression: *Class(User) && Inside_Package(SubscriberPackage)* will be tokenized by the scanner into three tokens (1) *Class(User)*, (2) *Inside_Package(SubscriberPackage)*, and (3) the logical operator &&. The parser will parse the textual expression and will translate it to the following OCL expression: *self.oclIsTypeOf(Class) and self.name='User' and self._package='SubscriberPackage'*. This expression will then be evaluated on the elements of the base model and the matched elements will be selected as join points.

5.3 Actual Weaving

During this step, the aspect adaptations are automatically woven into the base model at the identified locations according to the specification of the security solution. In the MOBS2 framework, we adopt a model-to-model transformation using the standard QVT (Query/View/Transformation) language. Model transformation is the process of generating target model(s) from source model(s). The input models of the QVT transformation are the base model and the specialized aspect model, and the generated output model is the woven model.

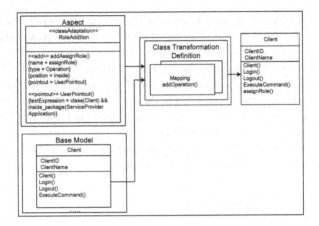

Fig. 9 Example of Class Transformation.

The weaving is specified as a set of transformation definitions. Each transformation definition consists of a set of mapping rules that specify how elements of the source model should be transformed to elements in the target model. In our weaving engine, we classify the transformation definitions according to the supported UML diagrams. Thus, we provide four types of transformation definitions: class transformation definition, sequence transformation definition, activity transformation definition, and state machine transformation definition. For instance, the *Class Transformation Definition* consists of a set of mapping rules which specify how each element of the class diagram is woven into the base model.

The actual weaving starts by parsing the adaptations specified in the aspect and according to the adaptation rules, the equivalent transformation definitions will be generated. Each adaptation rule will then be translated to QVT mapping rules. Figure 9 shows an aspect that contains a class adaptation where an add adaptation rule is specified to add an operation to a designated class. Having an adaptation of type class adaptation and the adaptation rule type set to Operation, the class transformation definition is going to be selected and the mapping rule *addOperation* will be executed. The result of this transformation will be the addition of the new operation *"assignRole()"* to the class *"Client"* of the base model.

6 Prototype

To demonstrate the feasibility of our approach, we have designed and implemented a prototype as a plug-in to the Rational Software Architect development environment [12]. The plug-in uses the AOM profile presented in Section 4 and provides the weaving capabilities needed to weave the aspects specified using the AOM profile into the base model. The plug-in provides also a graphical user interface to ease the specialization of aspects and their weaving in a systematic way. In this section, we show the feasibility of our approach by illustrating how the prototype can be

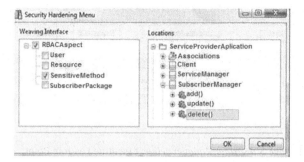

Fig. 10 Aspects Specialization.

used by the designer to apply the RBAC aspect (Figure 7) to the base model of the service provider application (Figures 2 and 3). The MOBS2 framework provides the developer with an RBAC aspect designed before hand by the security expert. This RBAC aspect is though application-independent and must be specialized by the developer to the service provider application.

6.1 Aspects Specialization through a Weaving Interface

As mentioned earlier, during this step, the developer specializes the application-independent aspect to his base model elements. To this end, we provide a graphical weaving interface that exposes only the generic elements that need to be instantiated (Figure 10). In our example, the developer maps *SensitiveMethod* to *Subscriber-Manager.delete()* (Figure 2). The same way, the developer maps *User* to *Client*, *Resource* to *Subscriber*, and *SubscriberPackage* to *ServiceProviderApplication*. Note that *RBACMonitor* in the RBAC aspect is used internally and transparently to the developer to monitor access rights from a client to any class with sensitive methods.

6.2 Join Points Identification

Having the aspect specialized to actual elements from the service provider application, each pointcut is automatically translated into its equivalent OCL expression using a parser generated by CUP parser generator for Java [3]. For example, the pointcut *SensitiveMethodPointcut* presented in Figure 7 with the textual expression: *"Message_Call(delete) && Message_Source(Client) && Message_Target(SubscriberManager)"* will be translated to the following OCL expression: *"self.oclIsTypeOf(Message) and self.name='delete' and self.connector._end->at(1).role.name='Client' and self.connector._end-> at(2).role.name='Subscriber-Manager'"*. This expression will then be evaluated on the base model and the matched elements will be selected as join points. Figure 11 shows the result of evaluating the previous OCL expression on the *DeleteSubscriber* sequence diagram.

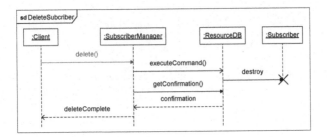

Fig. 11 The Message *SubscriberManager.delete()* Identified as Join Point.

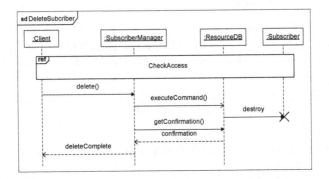

Fig. 12 Woven Model of DeleteSubscriber.

6.3 Actual Weaving

After identifying all the existing join points, the next step will be to inject the different adaptations into the exact locations in the base model. This is done by executing the QVT mapping rules that correspond to the adaptation rules specified by the security expert. We implemented the mapping rules using the Eclipse M2M QVT Operational plug-in [5] on top of RSA. These mapping rules are then interpreted by the QVT transformation engine that transforms the base model into a woven model. Figure 12 shows the final result after weaving the *CheckAccess* fragment of Figure 7 into the *DeleteSubscriber* sequence diagram of Figure 3.

7 Related Work

In the following, we first present the related work on AOM for security before discussing the previous work on design weaving. Regarding the specification of aspects in UML, considerable work has been done in the literature. An overview of existing AOM contributions can be found in [17, 19]. However, most of existing approaches are programming language dependent and only cover few of AOP capabilities.

Regarding the use of AOM for security, many approaches have been published recently [23,7,16,9,10,22,4]. [4] proposes an aspect-oriented framework to support the design and analysis of non-functional requirements defined as reusable aspects for distributed real-time systems using UML and formal methods. [7] uses UML diagram templates for modeling access control aspects as patterns. [9] proposes an aspect-oriented design approach for CORBA AC; a reference model for enforcing access control in middleware applications. [10] proposes an aspect-oriented risk-driven methodology for designing secure applications. After evaluating the application against defined attacks, and if the application presents a security risk, then a security mechanism specified as an aspect is incorporated into the application. [16] extends the UML meta-model with new diagrams to represent access control requirements. [22] proposes an approach for the analysis of the performance effects of security properties specified as aspects. [23] extends the UML-based Web Engineering method by specifying the detailed behavior of navigation nodes using state machines. Access control to navigation nodes is specified by refining the default state machines by a state machine modeling the access control rules.

Various approaches have been proposed for weaving UML design models. [24] presents Motorola WEAVR; a tool for weaving aspects into executable UML state machines. However, this weaver is based on the Telelogic TAU G2 implementation, therefore, it is tool-dependent and not portable. [8] proposes a model weaver for aspect-oriented executable UML models. The considered join point model intercepts only the interaction between objects. [11] presents XWeave for weaving models and meta-models. XWeave is based on the Eclipse Modeling Framework. [6] presents a generic tool for model composition called Kompose that is built on top of Kermeta [15]. It focuses only on the structural composition of any modeling language described by a meta-model.

8 Conclusion and Future Work

This paper presents an approach for specifying and weaving security aspects into UML design models. For the specification of aspects, we designed a UML profile allowing the specification of common aspect-oriented primitives. In addition, we defined a UML-specific pointcut language to designate UML join points. Furthermore, we developed a framework to automatically weave the aspects into base models. By adopting the standard OCL for evaluating the pointcuts, our approach is generic enough to specify a wide set of pointcut expressions covering various UML diagrams. The adoption of the standard QVT for implementing the adaptation rules extends portability of the designed weaver to all tools supporting QVT beyond current implementation in RSA. In the future, to complete the full life cycle of the software, we will investigate the generation of secure code from the woven models.

References

1. AOM Website: http://www.aspect-modeling.org/
2. Bodkin, R.: Enterprise Security Aspects. In: Proc. of the 4th Workshop on AOSD Technology for Application-Level Security (2004)

3. CUP Parser Generator for Java, `http://www2.cs.tum.edu/projects/cup/`

4. Dai, L., Cooper, K.: Modeling and Analysis of Non-Functional Requirements as Aspects in a UML Based Architecture Design. In: Proc. of the Sixth Intl. Conference on Software Engineering, Artificial Intelligence, Networking and Parallel/Distributed Computing, pp. 178–183. IEEE Computer Society, Washington (2005)

5. Dvorak, R.: Model Transformation with Operational QVT (2008), `http://www.eclipse.org/m2m/qvto/doc/M2M-QVTO.pdf`

6. Fleurey, F., Baudry, B., France, R., Ghosh, S.: A Generic Approach for Automatic Model Composition. In: Proc. of 11th Intl. Workshop on AOM, pp. 7–15. Springer, Nashville (2007)

7. France, R., Ray, I., Georg, G., Ghosh, S.: AO Approach to Early Design Modelling. Software, IEE Proceedings 151(4), 173–185 (2004)

8. Fuentes, L., Sánchez, P.: Designing and Weaving AO Executable UML Models. Journal of Object Technology 6(7), 109–136 (2007)

9. Gao, S., Deng, Y., Yu, H., He, X., Beznosov, K., Cooper, K.: Applying Aspect-Orientation in Designing Security Systems: A Case Study. In: Proc. of the Intl. Conference of Software Engineering and Knowledge Engineering (2004)

10. Georg, G., Houmb, S.H., Ray, I.: Aspect-Oriented Risk-Driven Development of Secure Applications. In: Damiani, E., Liu, P. (eds.) Data and Applications Security 2006. LNCS, vol. 4127, pp. 282–296. Springer, Heidelberg (2006)

11. Groher, I., Voelter, M.: XWeave: Models and Aspects in Concert. In: Proc. of the 10th Workshop on AOM, pp. 35–40 (2007)

12. IBM-Rational Software Architect, `http://www.ibm.com/software/awdtools/architect/swarchitect/`

13. Miller, J., Mukerji, J.: MDA Guide Version 1.0.1. Tech. rep., Object Management Group (OMG) (2003)

14. Mouheb, D., Talhi, C., Lima, V., Debbabi, M., Wang, L., Pourzandi, M.: Weaving Security Aspects into UML 2.0 Design Models. In: Proc. of the 13th Workshop on Aspect-Oriented Modeling, pp. 7–12. ACM, New York (2009)

15. Muller, P.A., Fleurey, F., Jézéquel, J.M.: Weaving Executability into Object-Oriented Meta-Languages. In: Briand, S.K.L. (ed.) MODELS/UML 2005. LNCS, vol. 3713, pp. 264–278. Springer, Heidelberg (2005)

16. Pavlich-Mariscal, J., Michel, L., Demurjian, S.: Enhancing UML to Model Custom Security Aspects. In: Proc. of the 11th Workshop on Aspect-Oriented Modeling (2007)

17. Chitchyan, R., et al.: Survey of Analysis and Design Approaches. Technical Report-AOSD-Europe-ULANC-9 (2005)

18. Sandhu, R., Ferraiolo, D., Kuhn, R.: The NIST Model for Role-Based Access Control: Towards A Unified Standard. In: Proc. of the fifth ACM workshop on Role-Based Access Control, pp. 47–63 (2000)

19. Schauerhuber, A., Schwinger, W., Kapsammer, E., Retschitzegger, W., Wimmer, M., Kappel, G.: A Survey on Aspect-Oriented Modeling Approaches. Technical Report, Vienna University of Technology (2007)

20. Viega, J., Bloch, J.T., Chandra, P.: Applying Aspect-Oriented Programming to Security. Cutter IT Journal 14, 31–39 (2001)

21. Win, B.D.: Engineering Application Level Security through Aspect-Oriented Software Development. PhD Thesis, Katholieke Universiteit Leuven (2004)

22. Woodside, M., Petriu, D.C., Petriu, D.B., Xu, J., Israr, T., Georg, G., France, R., Bieman, J.M., Houmb, S.H., Jürjens, J.: Performance Analysis of Security Aspects by Weaving Scenarios Extracted from UML Models. Journal of Systems and Software 82(1), 56–74 (2009)
23. Zhang, G., Baumeister, H., Koch, N., Knapp, A.: AO Modeling of Access Control in Web Applications. In: Proc. of the 6th Workshop on Aspect-Oriented Modeling (2005)
24. Zhang, J., Cottenier, T., Berg, A., Gray, J.: Aspect Composition in the Motorola Aspect-Oriented Modeling Weaver. Journal of Object Technology. Special Issue on AOM 6(7), 89–108 (2007)

Study of One Dimensional Molecular Properties Using Python

Eric. O. Famutimi, Michael Stinson, and Roger Lee

Summary. One of the attractions to the study of one dimensional systems is the technological interest of their possible effects in nanoelectronics [1]. There are myriads or papers on the solution to the problem of the electronic properties of one dimensional systems. Few of these papers use python for visualization but none has used python as a tool for solving this problem from first principle. In this paper, we present several techniques of using Python as a tool in computational analysis. We report the results of using python to study the electronic properties of an infinite linear chain of atoms. We use the principles of nearest neighbor and directly calculated the eigenvalues of our system. We also derived the green function for the system and compared the eigenvalues obtained from the green function with those directly calculated. Visualization of our results was achieved using Matplotlib, a powerful yet, easy to use Python plotting library. Our results show an agreement between the eigenvalues obtained by direct calculation and those obtained using our derived green function for the system. The results also show the simplicity of Python as an analytical tool in computational sciences.

1 Introduction

From scientific software developer point of view, one of the biggest obstacles is developing software that is easy to use, simple to maintain, adaptable and reusable. One of the major advantages of having these qualities is the ability of the user to spend more time doing science that writing codes. Using a chain of linear molecules as a case study, this paper demonstrates how python could be used as an effective tool for development of scientific software.

With the quest for miniaturization of electronic components, the study of electronic properties of materials is one of the fastest growing research areas in the field of material science. In 1936 Johannson and Linde discovered the existence of a long-period superlattice in CuAu II [2]. Since then, there had been considerable interest in the study of electronic properties of such molecular topologies. The interest in the study of these one dimensional systems is on the one hand, due to their unique fundamental properties as shown by Lu et.al. in their

Eric. O. Famutimi, Michael Stinson, and Roger Lee
Department of Computer Science Central Michigan University,
Mount Pleasant MI, USA
e-mail: `famutleo@cmich.edu`, `stinslm@cmich.edu`, `leelry@cmich.edu`

Roger Lee (Ed.): SERA 2010, SCI 296, pp. 215–226, 2010.
springerlink.com   © Springer-Verlag Berlin Heidelberg 2010

extensive analytical and numerical study of the properties of one-dimensional quasillattices [3]. Nui and Nori applied the principles of renormalization group approach to study the properties of one dimensional quasiperiodic systems [4]. On the other hand, one dimensional electronic structures have peculiar advantages over their bulk counterparts in nanoelectronics applications. Xia et al. gave a comprehensive review on the synthesis, characterization and applications of one dimensional nanostructure [5].

Several tools for performing numerical analysis have been created. These tools are developed using traditional programming languages such as Fortran, C and C++. Within the community of computational scientists, FORTRAN seems to be the most popular language. This is however not surprising giving the fact that the first version of FORTRAN was released in 1954. There are therefore mammoth of library functions contributed by computational scientists that have used this programming language since its inception. During the last decade however, computational scientists have gradually transitioned into using other computing environment such as IDL, Maple, Mathematical, and in the recent few years; Python. This transition is occasioned by the simple and clear syntax of the command languages in these environments. This is coupled with a powerful and tight integration of simulation and visualization tools that are built into languages like Python. Calls to these visualization tools can be initiated with few lines or in some cases one line of code. Recent developments have shown that Python is more elegant, robust and user friendlier. It allows users to build own environment tailored to specific needs and integrated into any code written in other high performance languages like Fortran, C or C++ [6].

The purpose of this paper is to elucidate the electronic properties of one-dimensional aperiodic systems using a programming language that was hitherto not popular for computational analysis. We calculated from first principle, the eigenvalues, Green's function, local density of state and the total density of state for a generalized linear system. Unlike previous papers that attempt to solve similar problems using Python, while such papers make calls to FORTRAN modules for numerical analysis and import outputs into Python for visualization, our paper made use of codes written in Python for both numerical analysis and visualization.

In the next section, we discuss related work on the study of the electronic properties of one dimensional systems. Section 3 describes the physics and the mathematical formulations that form the basis of our analysis. In section 4 we present a step by step formulation of our problem and the methodology adopted in achieving results. Section 5 discusses the results obtained and section 6 summarizes future area of research based on our results in this paper.

2   Review of Related Work

The electronic properties of one dimensional quasiperiodic systems have been extensively studied [8-10]. Studies of this nature could either be experimental or computational. Computational analysis could either be analytic or numerical. Method employed for investigation is highly dependent on the problem at hand

and the defined goals and objectives of the research. However, irrespective of the method used, complexity of software is one of the greatest single challenges facing modern high performance scientific computing [11]. It is therefore imperative that scientific software have to evolve with this ever increasing complexity and afford the scientist the opportunity to do science rather than modifying codes to meet their needs.

2.1 Experimental Approach

There are various methods and techniques employed in experimental studies of the electronic properties of materials. Holzmann et al. studied one dimensional transport of electrons in Si/Si0.7Ge0.3 heterostructures [12]. They used a combination of laser holography and reactive ion etching techniques to study magnetotransport of high-mobility electrons in quasi-one-dimensional Si/Si0.7Ge0.3. In their study, Typical features of transport in narrow electron channels, such as oscillations due to the depopulation of quasi-one-dimensional subbands and an anomalous resistance maximum at low magnetic fields were observed.

Experimental methods give insight into molecular properties of materials by direct measurement. This is very useful when materials are synthesized and they have to be characterized. For example, Berger et al. studied the electronic properties of Nb4Te17I4 that was synthesized in the laboratory. They confirmed through resistivity measurements that this compound exhibits a pseudo-one-dimensional behavior. Through photoemission spectroscopy, they were able to clarify its valence band states [1]. From their measurements and analysis, they concluded that this compound is a narrow-gap semiconductor at room temperature, with a forbidden gap width of 0.34 eV.

It is a common knowledge that the purity of some valued metals re measured by the amount of impurities embedded in them. In a similar manner, the electronic properties of nanostructures consisting of only a few atoms may depend sensitively on the exact position and chemical identity of each and every atom in the structure [13]. As shown in [14], a single atom impurity could considerably alter the transport properties of atomic scale conductors. Wallis et al. used scanning tunneling microscope to study artificial Au atomic chains with individual Pd impurities that were assembled from single metal atoms [13]. Wallis and his group further showed various electronic resonances of Au for varying sizes of Pd. Using conductance images, they revealed delocalized electronic density oscillations in pure Au segments.

2.2 Computational Approach

One of the beauties of computational methods is the ability to solve physical problems without expending resources in the purchase of specialized equipments. All you need is an access to a computer and analysis software for computation.

We are not by any means advocating that computational method is better than experimental methods or vice versa. These two methods complement each other.

Bose et al considered a one-dimensional superlattice consisting of two different square well potentials of finite strength and size. By using the transfer matrix formalism, they showed that it is possible to obtain exact solutions for all states and for all sizes of the antiphase domains. They also calculated the density of states of the electrons in both the bound and the ionization energy ranges [15].

Avishai and Band derived the relationship between the one-dimensional density of state at energy E and the phase of the transmission amplitude at that energy [16]. They developed a two-channel S-matrix scattering theory for one-dimensional scaterring and employing the relationship between the trace of the Green's function operator and the on-shell S matrix. When this is related to our calculations, the density of state calculated by Avishai and Band are actually the total density of state which is a function of the trace of the Hamiltonian matrix we obtained for our chain of linear atoms.

Using analytical methods involving continued fractions, Newman [17] calculated the site-site, two-time Green functions for a general one-dimensional crystal. The Green's functions as calculated give the electron correlations for a tight-binding model in the non interacting electron approximation, or the displacement correlations of a vibrating system with harmonic interatomic forces. He used the formulas generated to determine the density of states of a large class of quasiperiodic models and checked the results obtained with against perturbations-theory calculations of the band structure of the model he adopted for his calculations. In our study, we will be using computational approach rather than analytical methods. It will however be instructive to compare our results with that obtained using analytical approach used by Newman.

2.3   Python as a Driving Tool

Most available literatures demonstrate how Python can be used as a driving tool for steering molecular simulations [18,19,20,21]. The understanding of very large molecular simulations is more than just a simulation problem, an analysis problem, a visualization problem or a user-interface problem but a combination of all these. Therefore, best solutions would only be achieved when all these are combined in a balanced manner [18]. Beazley and Lomdahl [18] demonstrated this when they used Python scripts to steer molecular simulations from SPaSM (Scalable Parallel Short-range Molecular Dynamics), a molecular simulation code written in C language. In this work and a related one [19], they showed how python scripts where used not only for rendering and remote visualization but also for managing the large data generated during computation.

In [19], Beazley and Lomdahl showed how they used a Simplified Wrapper and Interface Generator (SWIG) to integrate C and Python codes. C libraries are compiled into shared libraries and are dynamically loaded into Python as extension modules. The Molecular Modeling Toolkit (MMTK) is a library of Python modules and C extensions. Hinsen [20] compared the performance of MMTK and CHARMM, a modeling program written in Fortran 77. When both, packages were applied to the equilibration of 300 water molecules with periodic boundary conditions, CHARMM was 2.5 times faster on a PentiumPro computer under Linux using gcc and g77 for compilation. However, while the CHARMM input file contain twenty statements, worked only after an hour of trial-and-error modification and required manual editing of the input geometry file, MMTK consists of only seven statements and worked correctly the first time without manual modifications to the input files.

In [21] Beazley and Lomdahl used python scripts to show how molecular simulations could be made to be more interactive. They provided various script samples and showed the comparison of some simple C and Python implementations.

3 Methodology

3.1 Mathematical Model

In the model adopted for this work, we represent the location of each atom by a node. Each atom has an onsite energy E and an interaction t. The interactions are restricted to nearest neighbor interactions such that atom n can only see atom $n \pm 1$. For this one dimensional model, we solve the Schrödinger equation:

$$-\frac{\hbar}{2m}\frac{d^2}{dx^2}\psi(x) + V(x)\psi(x) = E\psi(x) \tag{1}$$

The solution to (1) for several forms of potential $V(x)$ reveals hidden information about the representative system. Let us assume our atoms are identical, and each centered at lattice position t_n along the x axis. If for any non degenerate local orbital ϕ_a we assign energy E_a. If H is the Hamiltonian of the system, from translational symmetry, the diagonal matrix elements (E_o) of H on each atomic orbital are all equal. In a similar manner, the hopping integrals (γ) between nearest neighbor orbitals are also equal. From above, it therefore follows that:

$$<\Phi_a(x - t_n)|H|\Phi_a(x - t_n)> = E_0 \tag{2}$$

$$<\Phi_a(x - t_n)|H|\Phi_a(x - t_{n\pm1})> = \gamma \tag{3}$$

3.1.1 Green's Function

For an arbitrary operator H, Green's function is defined as

$$G_{00}(E) = \frac{1}{E - H} \tag{4}$$

Given a state of interest $|f_o>$ normalized to 1 and a model crystal Hamiltonian H, the Green's function $G_{00}(E)$ is defined as

$$G_{00}(E) = \triangleleft f_0 \frac{1}{E-H} | f_0 > \tag{5}$$

In practice, the real energy E is accompanied by an infinitesimal positive imaginary part ε to take care of poles at certain values of E. It therefore follows that in practice, Green's function as used in our analysis is defined as

$$G_{00}(E) = \triangleleft f_0 \frac{1}{E + i\varepsilon - H} | f_0 > \tag{6}$$

On the real energy axis, the real part of $G_{00}(E)$ exhibits poles corresponding to the discrete eigenvalues of H. In our analysis, we will be calculating these eigenvalues and compare them to those obtained by direct matrix calculations.

3.2 Model Setup

The goal of this study is to explore the applicability of Python as a tool for numerical analysis in Physics. We chose as a test model, a system of infinite linear chain of nodes. Each node has same onsite energy E and interactions t between sites A_n and A_{n+1}. In other words, we used the nearest neighbor tight biding approach.

This model could be physically represented by a linear chain of homogenous atoms or molecules with an interaction negligible enough that it only affects the next neighboring node. A simple representation of this model is as shown in figure 4.1 below. $\varphi(n)$ defines the state of atom A_n.

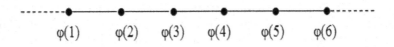

$$\varphi(1) \qquad \varphi(2) \qquad \varphi(3) \qquad \varphi(4) \qquad \varphi(5) \qquad \varphi(6)$$

Fig. 3.1 A Schematic Representation of our model.

3.3 Calculations

In achieving our set objectives, we methodically did the following.

- We set up a generalized one dimensional problem and wrote python codes that set up the Hamiltonian of the system by reading the number of site, the onsite energy and the nearest neighbor interaction from a file. This part of the analysis uses four functions and the output is a tridiagonal matrix with rank equal to the number of atoms and diagonal elements equal to the onsite energies.The next step in our analysis was to calculate the eigenvalues using standard built in functions from numpy. Numpy is a library functions written in Python for matrix operations and some other numerical operations especially on matrixes. These eigenvalues are compared to those obtained from our Green's function.

- We used the Hamiltonian matrix obtained in previous step to generate the Green function matrix elements. As shown in equation 6, it should be noted that the addition of the scalar quantity $E + i\varepsilon$ to the operator H in matrix form implicitly implied that the scalar quantity is preliminary multiplied by the identity operator. We achieved this in Python by using the built in function eye(m) which generates a square identity matrix with rank m.
- We generated the Green function for the outtermost atom by calling $G_{00}(E + i\varepsilon)$. Mathplotlib was used to display the result of the variation of $G_{00}(E + i\varepsilon)$ with E. We also calculated the Green function for other sites by calling Gii where $i = 0 \rightarrow n-1$ and n is the number of sites.

4 Results and Discussion

In our presentation of results, we will present results for a linear chain of six nodes. We will show the Hamiltonian generated by our code and print the result of the calculated eigenvalues. We will also show the plots of the Green's function obtained and show the behavior of this function around the poles to reveal the eigenvalues obtained using Green's function.

4.1 Hamiltonian Matrix

The generated Hamiltonian is as shown in equation 4.1.

$$\begin{pmatrix} 0 & -0.1 & 0 & 0 & 0 & 0 \\ -0.1 & 0 & -0.1 & 0 & 0 & 0 \\ 0 & -0.1 & 0 & -0.1 & 0 & 0 \\ 0 & 0 & -0.1 & 0 & -0.1 & 0 \\ 0 & 0 & 0 & -0.1 & 0 & -0.1 \\ 0 & -0.1 & 0 & 0 & -0.1 & 0 \end{pmatrix} \tag{4.1}$$

From inspecting equation 4.1, it can be seen that the Hamiltonian obtained is a tridiagonal matrix with diagonal elements A_n whose values are zero while the $A_{n\pm1}$ elements have values -0.1. This is the expected Hamitonian as explained in our model. The onsite energies have values equal to zero while the nearest neighbor interaction is taken to be negative 0.1.

4.2 Eigenvalues

The calculated eigenvalues are as shown in table 4.1.

Table 4.1 Table of Eigenvalues

Site number	Eigenvalue
1	-0.18019377
2	-0.12469796
3	-0.04450419
4	0.04450419
5	0.12469796
6	0.18019377

From table 4.1 above, we can see the symmetry in the calculated eigenvalues. These calculated values would be compared to those obtained when we use Green functions to calculate the eigenvalues of the system.

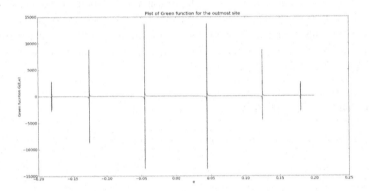

Fig. 4.1 Green plot for G00

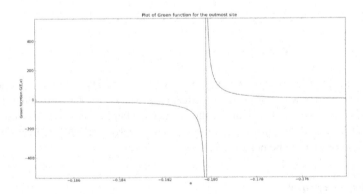

Fig. 4.2 Green plot revealing pole at site 1

4.3 Plots of Green's Function

The plots obtained from the generated Green's functions are as shown in fig 4.1 to 4.7. Figure 4.1 is the plot of the Green function as a function of E for all the six sites in consideration.

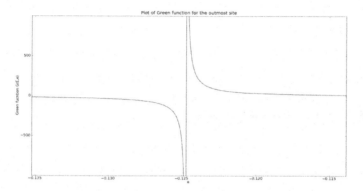

Fig. 4.3 Green plot revealing pole at site 2

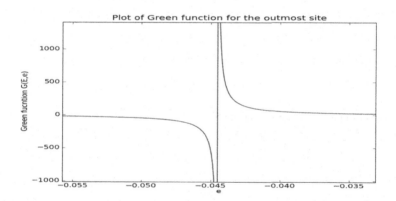

Fig. 4.4 Green plot revealing pole at site 3

As shown in figures 4.2 to 4.7, it can be seen that the poles shown in the figures correspond to distinct eigenvalues shown in Table 4.1. This confirms that our calculated Green Function is correct and produced the expected results.

4.4 Comparison with Fortran

Rather than directly calling specialized functions and routines in codes written in Fortran, python has library functions that do all the backend calculations, presenting you with well formatted results.

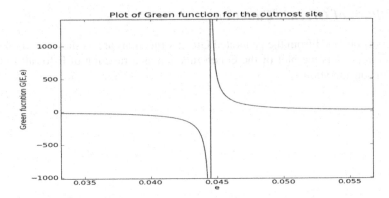

Fig. 4.5 Green plot revealing pole at site 4

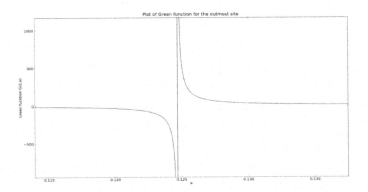

Fig. 4.6 Green plot revealing pole at site 5

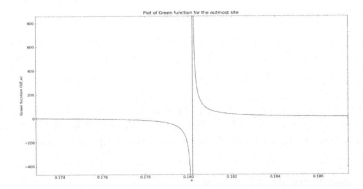

Fig. 4.7 Green plot revealing pole at site 6

As a typical example, In the LAPACK package, a call to the subroutine zgeev that calculates eigenvalues and eigenvectors requires fourteen input parameters. This same calculation can be done either interactively or as a script in python with a single call and a single parameter. During execution, even though Python makes a call to say zgeev, it simplified the call to this subroutine by providing separate library functions for each parameter of interest. For example, eig(A) gives the eigenvalues and eigenvectors of matrix A while eigvals(A) gives only eigvalues of A. This affords scientist the flexibility and the extra time needed to focus on primary problem of interest which is science.

5 Suggestion for Future Work

In this study, we calculated the eigenvalues and generated a Green function for our model. We can further this by using the generated Green's function to calculate the local density of state and the total density of state for the system in consideration. This is a task we plan to embark on in the next phase of this work.

Acknowledgment. This work is supported by grant from the Department of Computer Science, Central Michigan University under the graduate assistantship program.

References

1. Berger, H., Marsi, M., Margaritondo, G., Forro, L., Beeli, C., Crotti, C., Zacchigna, M., Comicioli, C., Astaldi, C., Prince, K.: On the electronic properties of the quasi-one-dimensional crystal $Nb_4Te_{17}I_4$. Journal of Physics D: Applied Physics 29(3) (1996)
2. Lu, J.P., Birman, J.L.: Electronic Structure of a Quasiperiodic System. Phys. Rev. B Rapid. Comm. 36, 4471–4474 (1987)
3. Lu, J.P., Odagaki, T., Birman, J.L.: Properties of One Dimensional Quasilattices. Phys. Rev. B 33, 4809–4816 (1986)
4. Niu, Q., Nori, F.: Renormalization-Group Study of One-Dimensional Quasiperiodic Systems. Phys. Rev. Lett. 57, 2057–2060 (1986)
5. Xia, Y., Yang, P., Sun, Y., Wu, Y., Mayers, B., Gates, B., Yin, Y., Kim, F., Yan, H.: One-Dimensional Nanostructures: Synthesis, Characterization, and Applications. Adv. Mater. 15(5) (2003)
6. Langtangen, H.P.: Python Scripting for Computational Science, 3rd edn. Springer, Heidelberg (2009)
7. Johansson, C.H., Linde, J.O.: Ann. Phys. (Leipzig) 25(1) (1936)
8. Kohmoto, M., Kadanoff, L., Tang, C.: Localization Problem in One Dimension: Mapping and Escape. Phys. Rev. Lett. 50, 1870 (1983)
9. Oallund, S., Pandit, R., Rand, D., Schellnhuber, H.J., Siggia, E.D.: One-Dimensional Schrödinger Equation with an Almost Periodic Potential. Phys. Rev. Lett. 50 (1983)
10. Kohmoto, M., Sutherland, B., Tang, C.: Critical wave functions and a Cantor-set spectrum of a one-dimensional quasicrystal model. Phys. Rev. B 35 (1987)
11. Bernholdt, D.E., Armstrong, R.C., Allan, B.A.: Managing complexity in modern high end scientific computing through component-based software engineering. In: Proceedings of HPCA Workshop on Productivity and Performance in High-End Computing (2004)

12. Holzmann, M., Tobben, D., Abstreiter, G., Wendel, M., Lorenz, H., Kotthaus, J.P., Schaffler, F.: One-dimensional transport of electrons in $Si/Si_{0.7}Ge_{0.3}$ heterostructures. Appl. Phys. Lett. 66(7) (1995)
13. Wallis, T.M., Nilius, N., Mikaelian, G., Ho, W.: Electronic properties of artificial Au chains with individual Pd impurities. J. Chem. Phys. 122, 011101 (2005)
14. Enomoto, S., Kurokawa, S., Sakai, A.: Quantized conductance in Au-Pd and Au-Ag alloy nanocontacts. Phys. Rev. B 65, 125410 (2002)
15. Bose, S.M., Feng, D.H., Gilnzore, R., Prutzer, S.: Electronic density of states of a one-dimensional superlattice. J. Phys. F: Metal Phys. 10, 1129–1133 (1980)
16. Avishai, Y., Band, Y.B.: One Dimensional Density of States and the Phase of the Transmission Amplitude. Phys. Rev. B 32 (1985) (Rapid Communication)
17. Newman, M.E.: Green's functions, density of states, and dynamic structure factor for a general one-dimensional quasicrystal. Phys. Rev. B 43(13) (1991)
18. Beazley, D.M., Lomdahl, P.S.: Lightweight Computational Steering of Very Large Scale Molecular Dynamics Simulations. In: Proceedings of the 1996 ACM/IEEE Conference on Supercomputing, p. 50 (1996)
19. Beazley, D.M., Lomdahl, P.S.: Feeding a large-scale physics application to Python. In: Proceedings of the 6th International Python Conference (1997)
20. Hinsen, K.: The Molecular Modeling Toolkit: a case study of a large scientific application in Python. In: Proceedings of the 6th International Python Conference (1997)
21. Beazley, D.M., Lomdahl, P.S.: Building Flexible Large-Scale Scientific Computing Applications with Scripting Languages. In: The 8th SIAM Conference on Parallel Processing for Scientific Computing (1997)

Comparing the Estimation Performance of the EPCU Model with the Expert Judgment Estimation Approach Using Data from Industry

Francisco Valdés and Alain Abran

Abstract. Software project estimates are more useful when made early in the project life cycle: this implies that these estimates are to be made in a highly uncertain environment with information that is vague and incomplete.

To tackle these challenges in practice, the estimation method most used at this early stage is the Expert Judgment Estimation approach. However, there are a number of problems with it, such as the fact that the expertise is specific to the people and not to the organization, and the fact that this intuitive estimation expertise is neither well described nor well understood; in addition, the expertise is difficult to assess and cannot be replicated systematically.

Estimation of Projects in Contexts of Uncertainty (EPCU) is an estimation method based on fuzzy logic that mimics the way experts make estimates. This paper describes the experiment designed and carried out to compare the performance of the EPCU model against the Expert Judgment Estimation approach using data from industry projects.

Keywords: EPCU, Estimation Projects, Uncertainty Contexts, Fuzzy Sets, Expert Judgment Estimation.

1 Introduction

1.1 Context

The information required for developing software is acquired progressively (i.e. based on the requirements) throughout the software development process, from the conceptualization phase to the final phase. The same applies to estimation: for instance, in the conceptualization phase, the inputs for estimation are derived from the rough information available at that time, at which point they are often expressed as variables with linguistic rather than numerical values, and with large

Francisco Valdés and Alain Abran
École de Technologie Supérieure
Dept. of Software Engineering
Montréal, Canada
e-mail: francisco.valdes@spingere.com.mx,
francisco.valdes.1@ens.etsmtl.ca, alain.abran@etsmtl.ca

Roger Lee (Ed.): SERA 2010, SCI 296, pp. 227–240, 2010.
springerlink.com   © Springer-Verlag Berlin Heidelberg 2010

ranges of imprecision and uncertainty. In practice, these large ranges of imprecision and uncertainty are most often managed in a somewhat *ad hoc* manner (intuitive approach), rather than with relevant specialized techniques.

On the one hand, the estimation approach using expert judgment in the early phases of a project has some benefits. For example, humans can manage a large amount of vague information simultaneously, using only linguistic values without quantitative details. However, treating vague information is a challenge in estimation models.

On the other hand, a number of problems arise when an intuitive approach based on expert judgment is used, because:

- it is founded on human expertise and past experience, and so is specific to the people and not to the organization,
- it is neither well described nor well understood, and
- it is hard to assess and cannot be replicated systematically.

In the software engineering field, a number of estimation models and tools have been developed over the past 40 years, focusing in particular on estimating project effort and, to a lesser extent, project duration. Software project estimates provided by (formal or informal) estimation models are used as a basis for decision making, and, in one way or another, influence the total cost of a project.

Most of the estimation models proposed to the industry can be classified in two major categories [3, 4]:

- algorithmic models
- non-algorithmic models

Algorithmic models are those most often found in the literature. Basically, these models are statistically derived from historical data on completed projects [5-7, 26]. A number of disadvantages of these models have been documented in [3, 4]:

- The prediction function form is pre-determined: for example, in the exponential model, $Effort = \alpha \ x \ size^{\beta}$, α represents the productivity coefficient and β the coefficient of economies/diseconomies of scale.
- Algorithmic models are obtained from contexts that may not be similar to those for which new projects have to be estimated. This makes adapting such models to the estimation context very challenging.

Non-algorithmic models [3, 4] attempt to model the complex relations between the independent variables (cost drivers) and the dependent variable (i.e. effort) more adequately. These models also have a number of disadvantages, such as:

- The knowledge required to implement any of them is very specific, and expert help is required most of the time. Such expert knowledge is difficult to describe and model, and is not readily available to practitioners.
- Usually, no easy-to-use end-user tools are available to implement these models, as they are typically developed with complex mathematical techniques and tools that are challenging for practitioners and managers to use.

- The knowledge required to identify analogy cases is very specific, and the computational effort required specifically for the analogy approach is considerable [9, 10, 11]

1.2 Handling Linguistic Values in Estimation Models

Most software estimation models proposed to the industry cannot accept linguistic values as inputs, since those models cannot interpret the imprecision and uncertainty embedded within them.

One way to deal with this limitation in software estimation models is to use fuzzy logic, which offers a well-structured quantitative framework for capturing the vagueness of human knowledge expressed in natural language [12]: in fuzzy logic, the expression of knowledge is (or is allowed to be) a matter of degree [12], which is how human thinking is organized [13].

There already exist a few studies in software measurement and estimation based on fuzzy logic [3, 4, 10, 14, 15, 16, and 17]. For instance, Idri *et al.* [3] proposed the Fuzzy Analogy Approach, which involves the fuzzification of the classical Analogy Approach and illustrated it with a fuzzification of the CO-COMO model [24].

More recently, Valdés *et al.* [16] proposed a software estimation model referred to as the Estimation of Projects in Contexts of Uncertainty (EPCU) model. This model takes into account:

- the linguistic variables used by practitioners to describe the input variables in their experience-based estimation process (when these inputs are based on the vague or ambiguous information available when they have to come up with a project estimate); and
- the way practitioners mix these linguistic values to arrive at a project estimate.

Such fuzzy logic-based estimation models are typically designed on the basis of the recognized expertise of experts who identify and intuitively quantify the inputs (i.e. the independent variables of the estimation model) with linguistic values and infer the value of the dependent variables (i.e. estimates of project effort and project duration).

The EPCU estimation model is designed to be used at the very early stages of software development. A key challenge is then to verify that such a model fulfills a key objective, which is *to improve the quality of the estimates made* a priori *by experts*.

This paper describes the experiment designed and carried out to compare the performance of the EPCU model against the Expert Judgment Estimation approach using data from industry projects. This paper is organized as follows. Section 2 presents the description of the EPCU model. Section 3 presents the experimental design. Section 4 presents the results obtained and the discussion. Finally, section 5 presents the conclusions.

2 EPCU Model

2.1 Overview of the Model

The EPCU model [16] has six steps: identification of the input variables, specification of the output variable, generation of the inference rules, fuzzification, inference rule evaluation, and defuzzification.

2.2 Identification of the Input Variables

The goal of this step is to elicit the most significant variables for a project or a kind of project from the experienced practitioners in an organization, along with an assessment of them: software size, software complexity, team skills, and so on.

It is natural for experts to have differing opinions about some variables. To deal with this, fuzzy logic is used in a step known as fuzzification, which is described in section 2.5.

Experienced practitioners must define the fuzzy sets for each variable, which means that they must classify the variables in terms of linguistic values which they can evaluate. For example, for the complexity parameter, the fuzzy set could be classified as low, average, or high.

Also required is to define the membership function domain to represent the opinions of the experienced practitioners about these input parameters. By the end of this step, the most significant parameters have been generated, together with their fuzzy sets and the ranges available for each of them.

2.3 Generation of Inference Rules

All the fuzzy sets belonging to each input variable must be combined into 'if…, then…' form:

If x and y, then z

If x or y, then z

where x is a fuzzy set for one input variable, y is a fuzzy set for another input variable, and z is the fuzzy set for the output variable.

All the fuzzy sets for each input variable must be combined to generate the rulebase.

2.4 Fuzzification

The goal of this step is to obtain fuzzified values as a consequence of opinions put forward by an experienced practitioner. This means that a membership function must be defined for the input variables.

If three fuzzy sets are used for the input variable, the membership function can look something like the example in Fig. 1.

Once the membership function is defined for all the input variables, an expert opinion needs to be requested for each variable. This process will create fuzzy values to be used in the next step to evaluate the rulebase.

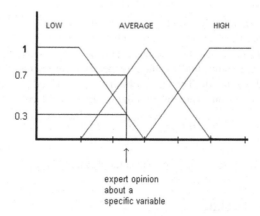

Fig. 1 Example of a fuzzy membership function and defuzzification

2.5 *Inference Rule Evaluation*

The fifth step consists in evaluating the rulebase by substituting the fuzzy values obtained for each input variable fuzzy set. The Inference Rule evaluation must follow the rules of fuzzy logic, such as:

$$Value\ (P\ \ or\ \ Q) = max\ \{value\ (P),\ value(Q)\}$$
$$Value\ (P\ and\ Q) = min\ \{value\ (P),\ value(Q)\ \}$$

2.6 *Defuzzification*

The sixth step is defuzzification, with the objective of obtaining a crisp value for the final estimate. Examples of defuzzification methods are: Max-Min, Max-Dot, Max-Product, Centroid Average, and Root Sum Square (RSS). The EPCU estimation generated in the case studies presented next was developed using RSS and then computing the 'fuzzy centroid' of the area.

This method was selected because it combines the effects of all applicable rules, scales the functions at their respective magnitudes, and computes the fuzzy centroid of the composite area. Even though it is more complex mathematically than the other methods, it was selected because it gives the best weighted influence to all the Inference Rules involved ("fired" in the specialized vocabulary).

The steps for obtaining the crisp value are:

1. Obtain the strength for each fuzzy set belonging to the output membership function (RSS). Considering the values obtained in the Inference Rule evaluation step, the strength for each fuzzy set defined for the output variable is evaluated with the following formula.

$$FS_k = (\Sigma R_i{}^2)^{0.5}$$

where FS_k, is the fuzzy set defined by a same linguistic value.

R_i is the rule that fired a specific fuzzy set.

2. Obtain the fuzzy centroid of the area. The weighted strengths of each output member function are multiplied by their respective output membership function center points and summed. The area obtained is divided by the sum of the weighted member function strengths, and the result is taken as the crisp output.

Crisp Value (FS_k) = Centroid = Σ ("FS_k" center * "FS_k" _strength) / Σ ("FS_k" _strength)

where FS_k, is the fuzzy set defined by the same linguistic value.

3 Experiment Design

3.1 *Experiment Participants*

Conducting the experiment reported here required that two different roles be played:

- that of the *expert*, who introduces the rules in the EPCU model for the context of each project to be estimated by the EPCU model. This is the individual from the organization providing the information about the projects who defines the context (input variables), the dependent variable to estimate (here, project duration), and the inference rules to be applied for estimating a specific project with the EPCU model.
- that of the *independent practitioners*, who are asked their opinion on the input variables for each of the projects to be estimated. The practitioners in this experiment were not familiar with the details of the software projects or with the organization's development contexts: the only basis that they had for evaluating the project inputs was their own experience as practitioners. Each of the practitioners participating in the experiment received the research questionnaire and had to evaluate the variables listed in it. They also had to provide their own (intuitive) estimate of project duration.

A total of 84 practitioners participated in this experiment. They were classified using the following three categories:

- their professional experience,
- their software development experience, and
- whether or not their profession is IT-related.

3.2   Experiment Phases

The experimental design consisted of 3 phases:

- A- Involvement of project experts for the development of the base material for the experiment, such as the descriptions of the input variables (i.e. the software requirements and the context of the projects);
- B- Involvement of 84 practitioners for selecting the values for the input variables for each of the projects to be estimated;
- C- Data analysis.

3.3   Phase A – Involvement of the Project Expert for Developing the Base Material for the Experiment

1. Selection of a set of completed projects

This step consisted of identifying a set of completed projects with the information necessary for this experiment. Five projects were selected. Information on 5 completed projects was available for this experiment, projects that were obtained from 4 distinct organizations. The information for building the EPCU model (i.e. providing the context information for each project) was provided by 4 experts, one from each of the 4 organizations (one expert provided the information for two projects from the same organization).

2. Description of the software requirements and the development context

In this step, the expert provides a description of the software requirements and also describes the context in which the project was developed.

A "context" is defined as a set of variables (input/output) and the relations that affect a specific project or a set of similar projects (i.e. rulebase).

Each of the 4 experts reproduced, for each distinct project, the corresponding information that was available in each organization at the time of the project's inception (that is, the independent variable as known at project's inception when *a priori* estimation is typically performed), even though the projects were actually concluded and the experts had all the information on the project.

For the purpose of this experiment in an *a priori* context, the participants were to be provided with the description of the software requirements as they were described in the early project phases. For this experiment, this required a re-documentation by the researcher of very early drafts of the preliminary statement of the scope of all the software to be developed. The re-documentation had to be performed at a very high level of abstraction, as is typically done by users at the conceptualization or feasibility stage in a software development process.

This re-documentation of the early software requirements was based on the availability of project documentation in each participating organization and in the experts' memories, as illustrated with the following example:

"The project is a .NET project to develop a B2B system for controlling the operations of shipping, transportation and delivery of packages for specialized organizations such as DHL or UPS. In addition, the B2B system must provide for contract and shipping management, package tracking, and so on."

The description of the context refers to the relevant factors that were present during project development. Most often, these factors are related to the process, the people, the organizational environment, and so on. For example:

"The responsibility of the entire project was assigned to a person who had a very good understanding about the problem domain, and of the tool in which the project was developed."

3.4 Phase B – Involvement of the Practitioners in Selecting the Input Values for Each of the Projects to Be Estimated

1. Collection of the practitioners' opinions about the input variables for each project

For this step, the practitioners were required to provide:

- a description of the software requirements (which was to be developed earlier in Phase A),
- a description of the development context in which the software was developed (as defined in Phase A), and
- A questionnaire designed to evaluate the input variables (the context variables) for each project.

Using the descriptions of the software requirements and the development context, the practitioners had to provide an opinion on each input variable (i.e. determine its value) for the 5 selected projects based on their own experience.

The information on the input variables was collected from the practitioners on three occasions between 2007 and 2008:

1. Electronic data collection: an email was sent to 28 people known to the researcher. Twelve (12) completed questionnaires were returned to the researcher.
2. Data collection within the context of a continuous education program at a university in Mexico for those in industry with a Master's degree in IT management. The questionnaires were distributed personally to 43 people, and 32 of them were returned.
3. Data collection during a November 2008 IT conference in Mexico on software measurement. At this conference, the questionnaire was distributed to approximately 150 people, and 40 completed questionnaires were returned.

For each input variable, the participants had to select values in a range of 0 and 5 (0 being the lowest and 5 the highest) [16].

2. Collection of the practitioners' expert judgment estimate for each project

In this step, the practitioners provided a duration estimate for each project (in months, weeks, and days) using the descriptions of the software requirements and of the development contexts, as well as their evaluation of these inputs and their own experience.

3.5   Phase C – Quality Criteria for Data Analysis

In the software engineering literature, a number of criteria are commonly used to evaluate the quality of estimation models [30], such as:

- the Magnitude of Relative Error (MRE).
  - o   Mean Magnitude of Relative Error (MMRE)
  - o   Median Magnitude of Relative Error (MdMRE)
- the Prediction level (Pred(l)). In the literature, an estimation model is considered good with Pred(25%) for 75% of the data points (or alternatively Pred(20%) for 80% of the data points).
- the Standard Deviation (SD_{MRE} = SDMRE = RMS)

The major advantage of the median MRE (MdMRE) over the mean MRE (MMRE) is that it is not sensitive to outliers, and therefore is more appropriate as a measure of the central tendency of a skewed distribution.

These quality criteria were used to compare the results generated by the EPCU model for the five projects selected for this experiment. For comparison purposes, it is relevant to consider the quality criteria with the dataset used, as well as with estimation models that use the same dataset.

Table 1 Performance Estimation results – full sample

	Project 1	Project 2	Project 3	Project 4	Project 5
MMRE	55%	16%	41%	34%	22%
MdMRE	65%	14%	37%	17%	24%
SD MRE	31%	11%	30%	31%	13%
Pred(25%)	21%	85%	42%	58%	80%

4   Data Analysis and Discussion

The descriptive statistics for the estimates gathered for the five projects were obtained using the SPSS 17 statistical software. The general overview is shown in Figure 2.

Analysis of the quality of the EPCU model
The quality of the use of the EPCU model, as evaluated with the 4 evaluation criteria for the 5 projects with the information from the 84 practitioners, is presented in Table 1. From Table 2, it can be observed that:

- the best MMRE (i.e. the lowest), obtained using the rulebase defined for estimating project 2, is 16%.
- the worst MMRE (i.e. the highest) is 55%.

If the MMRE is considered as the key criterion, the best rulebase performance is for project 2 (MMRE = 16%) and project 5 (MMRE = 22%), while projects 4 and 3 have an MMRE of 24% and 41% respectively.

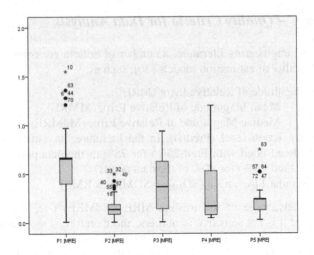

Fig. 2 Descriptive statistics on the duration estimates for the 5 projects

Considering next the coefficients of prediction better than 25%, the Pred(25%) criterion (bottom line of Table 1) is as follows:

- The most accurate estimation is obtained with the rulebase for project 2, with a prediction level of 85%, and project 5, with a prediction level of 80%.
- The least accurate estimation is obtained for project 1, with a prediction level of only 21%.

In order to evaluate the performance of the EPCU model against the practitioners' *a priori* estimates (i.e. an expert judgment-type estimation approach), the MMRE, Pred(25%), and the SDMRE obtained for the samples using both estimates (those generated by the EPCU model and the practitioners' estimates) were compared.

It must be noted that the information on the estimates of the practitioners using the Expert Judgment Estimation approach were collected only for the second and the third events (see Phase B – Involvement of the practitioners in selecting input values for each of the projects to be estimated), which is why the numbers of participants are less than 84. The results are presented in Table 2.

As can be seen in Table 2:

- MMRE criterion: the results obtained are considerably better (i.e. lower) for most projects (an improvement varying from 21% to 71%), with the exception of project 1 (with a decrease of 7% in this criterion). See Figure 3.
- SDMRE criterion: the results obtained for all projects by the EPCU model are much better (i.e. lower) than that obtained using Expert Judgment Estimation. The best improvement is for project 3: from an SDMRE of 112% for Expert Judgment Estimation, down to 30% for the EPCU model. See Figure 4.

Table 2 Comparison of results obtained using the EPCU model and Expert Judgment Estimation

		FULL SAMPLE using EPCU	Practitioner Judgment Estimation			FULL SAMPLE using EPCU	Practitioner Judgment Estimation
Number of practitioners		84	52	Number of practitioners		84	51
	MMRE	55%	48%		MMRE	16%	57%
	MdMRE	65%	50%		MdMRE	14%	67%
	SD MRE	31%	42%		SD MRE	11%	29%
P1	Pred(25%)	21%	33%	P2	Pred(25%)	85%	24%
Number of practitioners		84	52	Number of practitioners		83	53
	MMRE	41%	112%		MMRE	34%	55%
	MdMRE	37%	78%		MdMRE	17%	49%
	SD MRE	30%	112%		SD MRE	31%	45%
P3	Pred(25%)	42%	12%	P4	Pred(25%)	58%	17%
Number of practitioners		83	51				
	MMRE	22%	54%				
	MdMRE	24%	50%				
	SD MRE	13%	25%				
P5	Pred(25%)	80%	27%				

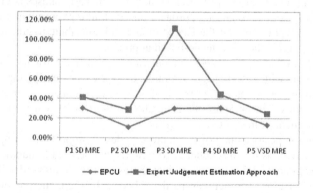

Fig. 3 MMRE comparison

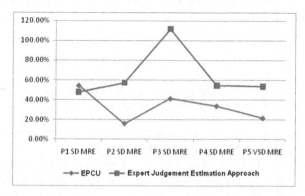

Fig. 4 SDMRE comparison

- Pred(25%) criterion: the results obtained with the EPCU model are considerably better (i.e. higher) for most projects (an improvement varying from 42% to 85%, as compared with 12% to 24% for the practitioners' estimates), with the exception of project 1 (with a decrease of 12% for this criterion).

5 Conclusion and Future Improvements

The EPCU fuzzy logic-based estimation model has been designed to be used in the early stages in the project life cycle. The objective of this study was to compare the performance of the EPCU model with the estimation method most frequently used in the industry for estimating software projects in the early phases (i.e. the Expert Judgment Estimation approach).

In the experiment reported here, the performance produced by the EPCU model for most of the projects is significantly better than that of the Expert Judgment Estimation approach, based on the quality criteria used. When the performance is better using the Expert Judgment Estimation approach, the difference is not too significant, so the performance can be considered equivalent.

Considering this, the use of the EPCU model in the early phases is preferable to the Expert Judgment Estimation approach, under similar experimental conditions. This illustrates that the use of the EPCU model can contribute to addressing some of the weaknesses of the experience-based approach.

The benefits of using the EPCU model can be significant for the industry: for this experiment, with information available from the very early stages, the estimates are similar, no matter what the skills of the people using the EPCU model. Furthermore, those who prepare the estimates are not fully involved in the subsequent development of the software projects.

Analyzing the performance of estimation models is important to improving the intuitive approach: as illustrated here, fuzzy logic can be used to handle the uncertainty and vagueness in software project estimation as humans do. The Expert Judgment Estimation approach can be described and stored, such as in the EPCU model, and can become part of an organization's assets. This constitutes a valuable solution for some of the problems described with the *ad hoc* experience-based estimation technique.

References

[1] Bourque, P., Oligny, S., Abran, A., Fournier, B.: Developing Project Duration Models in Software Engineering. Journal of Computer Science and Technology 22, 348–357 (2007)
[2] Project Management Institute, PMBOK Guide – A Guide to the Project Management Body of Knowledge, 3rd edn., p. 378 (2004)
[3] Idri, A., Abran, A., Khosgoftaar, T.M.: Fuzzy Analogy: A New Approach for Software Cost Estimation. In: International Workshop on Software Measurement (IWSM 2001), Montréal, Québec (2001)

[4] Idri, A., Abran, A., Khoshgoftaar, T.M., Robert, S.: Estimating Software Project Effort by Analogy Based on Linguistic Values. In: 8th IEEE International Software Metrics Symposium, Ottawa, Ontario (2002)

[5] Boehm, B.W.: Software Engineering Economics. Prentice Hall, Englewood Cliffs (1985)

[6] International Function Point User Group, http://www.ifpug.org/

[7] Ribu, Kirsten: Estimating Object-Oriented Software Projects with Use Cases, MSc thesis, University of Oslo, Department of Informatics (November 2001)

[8] Myrtveit, E.S.: A Controlled Experiment to Assess the Benefits of Estimating with Analogy and regression Models. IEEE Transaction on Software Engineering 25(4), 510–525 (1999)

[9] Shepperd, M., Schofield, C., Kitchenham, B.: Effort Estimation Using Analogy. In: ICSE-18, Berlin, pp. 170–178 (1996)

[10] Idri, A., Abran, A., Khoshgoftaar, T., Robert, S.: Fuzzy Case-Based Reasoning Models for Software Cost Estimation. In: Soft Computing in Software Engineering: Studies in Fuzziness and Soft Computing. Springer, Heidelberg (2004)

[11] Kolodner, J.L.: Case-Based Reasoning. Morgan Kaufmann, San Francisco (1993)

[12] Kacprzyk, J., Yager, R.R.: Emergency-oriented expert systems: a fuzzy approach. Information Sciences 37(1-3), 143–155 (1985); Referenced in [13]

[13] Azadeh, A., et al.: Design and implementation of a fuzzy expert system for performance assessment of an integrated health, safety, environment (HSE) and ergonomics system: The case of a gas refinery. Inform. Sci. (2008), doi:10.1016/j.ins.2008.06.026

[14] Idri, A., Abran, A.: Towards A Fuzzy Logic Based Measures for Software Projects Similarity. In: 6th MCSEAI 2000 – Maghrebian Conference on Computer Sciences, Fez, Morocco (2000)

[15] Shepperd, M., Schofield, C.: Estimating Software Project Effort Using Analogies. In: ICSE-18, Berlin, pp. 170–178 (1996)

[16] Souto, F.V., Abran, A.: Industry Case Studies of Estimation Models Using Fuzzy Sets. In: International Workshop on Software Measurement – IWSM-Mensura 2007, Palma de Mallorca (Spain), November 5-8, pp. 1–15 (2007)

[17] Kadoda, G.M., Cartwright, L.C., Shepperd, M.: Experiences Using Case-Based Reasoning to Predict Software Project Effort, EASE, Keele, UK (2000)

[18] Jeffery, R., Ruhe, M., Wieczorek, I.: Using public domain metrics to estimate software development effort. In: Seventh International Symposium on Software Metrics 2001, UK, pp. 16–27 (2001)

[19] IEEE Standard Glossary of Software Engineering Terminology, IEEE std. 610.12-1990 (1990)

[20] Condori-Fernandez, N., Pastor, O., Abran, A., Sellami, A.: Introduciendo Concep-tos de Metrologia en el Diseno de Medidas de Software. In: XI Iberamerico Workshop on Requirements Engineering and Environments, IDEAS 2008, Pernambuco, Brazil, pp. 112–125 (2008)

[21] The Standish Group International, Extreme Chaos, The Standish Group International, Inc., Research Reports (2000-2004), http://www.standishgroup.com

[22] The Buzz, Off Base: Insufficient expertise in setting baselines hits U.S federal IT budgets where it hurts. PM Network, 21 (March 2007)

[23] Zadeh, L.A.: Is there a need for fuzzy logic? Information Sciences 178(13), 2751–2779 (2008)

[24] Idri, A., Kjiri, L., Abran, A.: COCOMO Cost Model Using Fuzzy Logic. In: 7th International Conference on Fuzzy Theory &Technology, Atlantic City, New Jersey (2000)

[25] IFPUG, Function Point Counting Practices Manual, Version 4.2.1, International Function Points Users Group (2005)

[26] Zadeh, L.A.: Is there a need for fuzzy logic? In: Fuzzy Information Processing Society, NAFIPS 2008. Annual Meeting of the North American, May 19-22 (2008)

[27] Roychowdhury, S., Wang, B.-H.: Measuring inconsistency in fuzzy rules. In: Fuzzy Systems Proceedings, IEEE World Congress on Computational Intelligence, May 4-9, vol. 2, pp. 1020–1025 (1998)

[28] ISO, International Vocabulary of Basic and General Terms in Metrology, International Organization for Standardization, Switzerland, 2nd edn. (1993), ISBN 92-67-01075-1

[29] Steve, M.: Software Estimation: Demystifying the Black Art. Microsoft Press (2006)

[30] Abran, A.: Estimation and Quality Models Based on Functional Size with COSMIC – ISO 19761. Draft, ch.6 (April 2008)

Investigating the Capability of Agile Processes to Support Life-Science Regulations: The Case of XP and FDA Regulations with a Focus on Human Factor Requirements

Hossein Mehrfard, Heidar Pirzadeh, and Abdelwahab Hamou-Lhadj

Abstract. Recently, there has been a noticeable increase of attention to regulatory compliance. As a result, more and more organizations are required to comply with the laws and regulations that apply to their industry sector. An important aspect of these regulations is directly related to the way by which software systems, used by regulated companies, are built, tested, and maintained. While some of these regulations require from these systems to support a very specific set of requirements, others, the focus of this paper, are concerned with the process by which the system has been built. The Food and Drug Administration (FDA) regulations, for example, impose stringent requirements on the process by which software systems used in medical devices are developed. One particular focus of the FDA regulations is on having a user-centered approach for building software for medical devices through the use of well-known concepts in the area of human factor engineering. In this paper, we discuss these requirements in detail and show how Extreme Programming, an agile process, lacks the necessary practices to support them. We also propose an extension to XP, that if adopted, we believe it will address this particular need of the FDA regulations for medical device software.

Keywords: Regulatory Compliance, FDA Regulations, Extreme Programming, Software Engineering.

1 Introduction

For many regulated companies, regulatory compliance has become an important part of their business regardless of geography and industry sector. There are a number of factors behind the recent increase of attention to regulatory compliance including corporate scandals and the need for accountability, the reliance on

Hossein Mehrfard, Heidar Pirzadeh, and Abdelwahab Hamou-Lhadj
Department of Electrical and Computer Engineering,
Concordia University, Montreal, QC, Canada,
e-mail: {h_mehrfa,s_pirzad,abdelw}@ece.concordia.ca

Roger Lee (Ed.): SERA 2010, SCI 296, pp. 241–255, 2010.
springerlink.com

Information Technology (IT) solutions and the necessity to protect and secure sensitive information [1]. As a result, more and more authoritative rules (i.e., regulations, laws, standards, and guidelines) are introduced every year putting further constraints on the way companies are operated, managed, controlled, and governed. Examples of these authoritative rules include Sarbanes-Oxley Act (SOX), the Health Insurance Portability and Accountability Act (HIPAA), the Food and Drug Administration (FDA) regulations, etc. Many of these authoritative rules have a direct impact on the way software systems, used by regulated companies, are developed and maintained. For example, a data records management tool, used by a health institution which is required to comply with HIPAA, must support many data security features such as a password-based protection mechanism, different levels of data access control, and frequent backups and data reliability techniques.

The overall objective of our research is to study ways to help software companies cope with the increasing customer demand for software systems that satisfy a large set of regulatory rules. These rules impact software development in many ways. Some of them manifest themselves as functional requirements that need to be supported by the final product. Many other regulations such as life-science regulations, and particularly FDA (the focus of this paper), are concerned with the process by which the software has been built. They define a set of artefacts, which vary significantly in coverage and depth that the process needs to produce in order for the resulting system to be compliant. Producing such artefacts would normally be feasible if one adopts a traditional process. However, traditional processes come with their own set of challenges such as a lack of flexibility to react to changing requirements. Agile software processes, which are popular alternatives, favour flexible development mechanisms but suffer from lack of documentation [2, 3].

The objective of this paper is three-fold:

- Discuss the human factors requirements that FDA regulations impose on software processes for building software for medical devices.
- Discuss how agile processes, in particular XP, lack the necessary mechanisms to satisfy the FDA requirements for human factors requirements.
- Propose an extension to XP that can be used in software projects that require FDA compliance.

Organization of the paper: We review XP in Section 2. In Section 3, we propose a generic framework for extending software development methodologies to support life science regulations. In Section 4, we describe the application of our framework to the FDA medical device regulations against XP with the focus on Human Factors Engineering (HFE), which is one of the most important focuses of the FDA regulations. In Section 5, we discuss related work. Finally, in Section 6, we conclude the paper and show future directions.

2 Extreme Programming (XP)

Many practitioners consider XP as the symbol of the agile methodology. This is, perhaps, because it is one of the first agile processes that has been proposed. In general, XP consists of a set of individual practices that when put together can yield a successful software process [4].

As illustrated in Figure 1, XP starts with an Exploration phase. In collaboration with the programmer, the customer writes stories about the system features that he expects to be available for the first release. Programmer leads the customer in this process by raising specific questions (e.g., "is the story testable?") [5, 6]. The Planning phase prioritizes the collected stories based on their business values for the following small release. The required time and efforts for a release is estimated on a release plan in this phase [5].

At the beginning of developing a release, customer picks up stories based on the business values that he assigns to those stories. Then programmers break down those prioritized stories to a number of tasks and estimate the required time and efforts for each task. Based on this estimation and the structural stories, the story cards are reprioritized again to produce an iteration plan for whole iterations [6]. For the first iteration, it is important to choose the stories that mandate the system structure (architecture) for consequent iterations [5]. During each iteration, the specific set of selected stories are implemented by pair of programmers and tested by performing acceptance testing (functional testing). The iteration is not considered successfully implemented until it passes the acceptance test, which is normally written by the customer for verifying that the system functionalities satisfy the customer's needs. Moreover, during the productionizing phase, a set of additional performance and quality tests are conducted for the current release. Then, the approved release is documented and deployed to the customer [5, 6]. After deploying the first release to the customer, in the maintenance phase, the project should keep the release running for the customer by enhancing it and fixing its existing bugs while producing new iterations simultaneously [5].

As Table 1 shows, XP consists of a number of roles, a set of practices, and work products. In the table, each role is composed of a number of sub-roles. For instance, the role "XP programmer" could be broke down to "XP architect", "XP interaction designer", "XP implementer", "XP programmer", and "XP integrator".

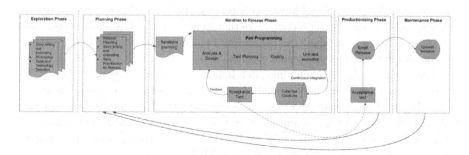

Fig. 1 XP Process Life Cycle (adapted from [5])

Table 1 The roles, practices and generated artefacts in XP

Role	Practice	Artefact
XP Coach	Adapt & improve process	None
	Explain process	
	Improve team skills	
	Keep process on track	
	Resolve conflicts	
XP Customer	Adjust iteration scope	User story
	Define customer test	XP customer test
	Define iteration	XP iteration plan
	Define release	XP release plan
	Report customer test result	XP vision
	Report project status	
	Revise release plan	
	Write user story	
XP programmer	Break down story	Coding standard
	Define coding standard	Metaphor
	Estimate task	Production code
	Estimate user story	XP build
	Implement spike	XP unit test
	Integrate & build	
	Refactor code	
	Write code	
	Help customer for story writing	
	Usage Evaluation	
	Large scale refactoring	
	System partitioning	
XP programmer(administrator)	Setup programmer environment	None
XP tester	Automate customer test	None
	Run customer test	
	Setup tester environment	
XP tracker	Track iteration progress	None
	Track release progress	

However, this classification is not rigid and like many other software processes, it could be characterized based on the requirements of the project at hand. In our approach, we used the main references of XP [5, 6, 7] and the Eclipse Process Framework (EPF) model library for XP [8] to provide this table.

3 A Framework to Extend Software Methodologies to Support Life-Science Regulations

Many Life-Science Regulations (LSRs) establish guidelines for software development in variety of products. External auditors (e.g., FDA auditors) seek for evidence that shows that the development team has complied with those guidelines during the development process.

LSRs often define the guidelines in a holistic way that is generic enough to be applied to various development methodologies. Unfortunately, this can cause lots of ambiguity for software developers as no specific development methodology can abide by the provided guidelines. For example, the FDA requests the medical device software developers to build safe and reliable software while no specific quality criteria are explicitly provided by the FDA.

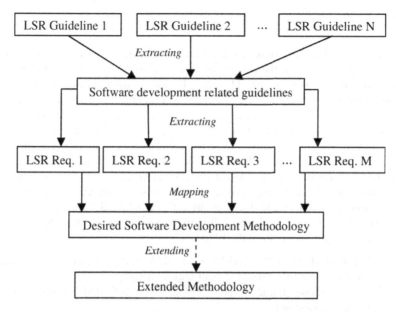

Fig. 2 A framework on how to meet life-science regulations for medical device software

Furthermore, LSRs often use terms that are not specific to software engineering. That is, a single term can be used in more than one field while having many completely different meanings. For instance, "risk analysis" can refer to an activity in both software requirements engineering and project management. This also can cause confusion in the intended meaning of a term making it difficult for development companies to comply with medical device guidelines as the developers do not know what they specifically have to follow.

In our approach, we alleviate the stated problems by following a framework through which we can extend a software development methodology of interest in a way that it can support a life science regulation. This framework (Figure 2) is composed of the following steps:

1. We visit the LSR guidelines and extract the guidelines related to software development.
2. We study these requirements from the software process engineering perspective and present typical software practices and documentation that can help developers follow the LSR guidelines.
3. According to the suggested practices and documentation for the LSR compliance, we investigate the capabilities of our desired software development methodology for supporting the LSR requirements.
4. Based on our evaluation on how well our desired software development methodology can support the extracted requirements, we propose a possible extension of the methodology to support the missing requirements.

4 FDA and XP

The Food and Drug Administration (FDA) regulations, is a LSR that imposes stringent requirements on the process by which software systems used in medical devices are developed [9, 10, 11, 12, 13]. These requirements translate into various software artefacts that must be made available for the software to be FDA-compliant. In this paper, we discuss these requirements in detail and show its possible lack of capability of an agile process such as XP to meet these requirements. For this we took the steps mentioned in our generic framework. First, we went through the FDA guidelines associated with medical devices and extracted the guidelines related to software development. Then, we studied these guidelines from the software engineering processes perspective and presented typical software practices and documentation that can help developers follow FDA guidelines. For this, we took the following three steps to extract the software development requirements from the FDA:

1. We went through the guidelines for FDA medical devices and extracted the related software development requirements.
2. Among those extracted requirements, we collected software process related requirements.
3. We clarified each of FDA software process requirements by proposing a set of software practices and documentations for each of them. This requirements clarification is done by looking at the FDA requirements from a software engineering perspective.

Table 2 shows the results of taking these steps. As illustrated in this table, we classified FDA software development requirements into four phases: Requirement, Design, Coding and Construction, and Testing. For each phase the required FDA practices are detailed.

Next, we investigated the capabilities of XP for supporting FDA requirements according to the suggested practices documentation for FDA compliance. For this, considering software process requirements that are extracted from FDA medical devices' guidelines, we evaluated XP capability for supporting these requirements. The result of this evaluation is reflected in Table 3.

As illustrated in Table 3, XP lacks support for many FDA practices and requirements. Therefore, we extended XP for the missing FDA requirements. Before starting to extend XP to support FDA requirements we tried to identify possible challenges. Due to direct or indirect interactions of medical devices' software with human lives, FDA requires many practices as well as documentations for verification and validation of developing software. The importance of documentation for FDA derives from achieving high quality software for developers and the fact that FDA auditors need some level of documentation to approve the software. The FDA software process requirements are more fitted to the type of software processes called plan-driven processes such as RUP. Plan-driven process is a disciplined process for software development that relies on heavily documented knowledge and stringent practices [14]. On the

Table 2 FDA requirements for medical device software

Required Validation - Software Analysis -	Methodology	Documentations
Requirements Elicitation	Interview, use case, observation and social analysis, focus group, brainstorming, joint application development (JAD), requirement modeling, prototyping.	Software Requirements Specification (SRS)
Requirements Evaluation	Formal review meetings, risk analysis, requirements inconsistency management, requirements prioritization, evaluation of alternative options in requirements, requirement verification, prototyping, and risk analysis.	Not specified.
Requirements Traceability Analysis	Not specified.	Software requirements and system requirements traceability matrix, software requirements and the risk analysis result traceability matrix
Test Plan	Not specified.	Acceptance test plan, system test plan
Configuration Management	Version control	Not specified.
Required Validation - Software Design -	**Typical Practices**	**Typical Documentations**
Human Factors Engineering	Usability testing and inspection, architecture sensitive usability patterns, scenario-based assessment of architecture.	Not specified.
Software Design Evaluation	Prototype evaluation, demonstration, simulation, comparing the design with other similar designs, analysis and inspection methods, compilation of relevant scientific literature, provision of historical evidence that similar designs are clinically safe.	Design review document, design verification document, design validation document, software design specification (SDS)
Design Traceability Analysis	Not specified.	Requirement and design traceability matrix
Update Test Plan	Not specified.	Module test plan, integration test plan
Test Design Generation	Not specified.	Test case, test procedure
Configuration Management	Version control	
Update risk analysis	Risk analysis of software design using techniques such as medical device use description.	Not specified.
Required Validation - Software Coding -	**Typical Practices**	**Typical Documentations**
Source Code Evaluation	Code audit, code inspection, code walkthrough, code review	
Source Code Documentation Evaluation	Not specified.	Source code document
Code Traceability Analysis	Not specified.	Traceability matrices for: source code to design specification, test cases to source code, test cases to design specification, test cases to risk analysis results, source code to risk analysis
Source code Interface Analysis	Interface checking	Not specified.
Test Generation	Not specified.	test case, test procedure
Required Validation Sub-Phases in Software Testing	**Typical Practices**	**Typical Documentations**
Test Documentation	Not specified.	Test plan, test procedure, test case, test report, test log.
Test Execution	Unit test, integration test, system test, user site testing	Not specified.
Test Traceability Analysis	Not specified.	Traceability matrices for: unit tests to detailed design, integration tests to high level design, and system tests to software requirements

other hand, XP emphasizes on less documentation and formality within the development life cycle. Thus, in extending XP we tried to reach a trade-off between keeping the process agile but at the same time inline with FDA. We extended XP by adding necessary sub-roles and practices that can support the requirements that were missing. Although we studied in detail FDA requirements and whether they are supported by XP or not, in this paper, we only discuss our

Table 3 Level of XP support for FDA requirements for medical device software

Development phases	Validation Sub-phases	XP Support for Typical Practices	XP Support for Typical Documentations
Requirements	Requirements Elicitation	User story card writing High customer involvement Eliciting requirements in number of iterations	No support for SRS
	Requirements Evaluation	System prototype Building software functionality in number of iterations Handling most probable risks early in development	Not enough documentation
	Requirements Traceability Analysis	No support	No documentation provided
	Test plan	Supported	Acceptance test plan No system test plan
	Configuration Management	No guideline	No documentation provided
Design	Human Factors Engineering	System prototype Not enough support	No documentation provided
	Software Design Evaluation	Informal design review Not enough support	No support for SDS and other documentations
	Design Traceability Analysis	No support	No documentation provided
	Update Test Plan	Supported	No documentation provided
	Test Design Generation	Supported	Unit test case Integration test case Acceptance test case No system test case
	Configuration Management	No guideline	No documentation provided
	Update risk analysis	No guideline	No documentation provided
Coding and Construction	Source Code Evaluation	Pair programming	No documentation provided
	Source Code Documentation Evaluation	No support	No documentation provided
	Code Traceability Analysis	No support	No documentation provided
	Source code Interface Analysis	No guideline	No documentation provided
	Test Generation	Supported	Unit test case Integration test case Acceptance test case No system test case
Testing	Test Documentation	Supported	Test Case No Support for the other documents
	Test Execution	Unit testing Integration testing Acceptance testing System testing	No documentation provided
	Test Traceability Analysis	No support	No support

mapping process between FDA and XP by focusing on one important FDA requirement, which consists of the need for a process to support Human Factors Engineering practices – This is important for medical software since any error can cause human lives.

4.1 FDA and HFE

The FDA highlights the importance of Human Factors Engineering (HFE) during the software design process. The FDA defines human factors engineering as "a discipline that should be taken during software and hardware design to improve human performance in using medical equipments". This improvement should be in accordance with end users' abilities [12].

Considering the human factors engineering during the system design can result in a product which causes fewer design-originated human errors. The FDA recognizes that the design for safety of medical devices should take into account human factors. The reason is that according to the FDA Center for Devices and Radiological Health (CDRH), the lack of attention to human factors during product development may lead to errors that can potentially cause serious patient injuries or even death [12].

The FDA medical device guidelines propose a set of requirements for HFE in medical device projects. In our study, we considered medical device projects as computer based system projects due to their software and hardware requirements.

In addition to HFE requirements for medical projects, FDA guidelines include HFE requirements during software development. The FDA highlights "software usability" as an HFE necessity to reach safety in software [9]. Furthermore, the FDA requires activities (e.g., usability test, risk analysis, prototyping and review) and produced documents (e.g., a test plan) during the software development process to accomplish the HFE requirement.

The FDA also provides a number of advices on how to deal with HFE on software design. To reach the HFE in software design, the FDA suggests "following Human Computer Interface (HCI) guidelines", "improving software usability", and "performing software design coordinated with hardware design". Here, we only consider the "improving software usability during software design" requirement as an example. To improve software usability, the FDA suggests a number of usability tests such as scenario-based testing, and testing the product by users per iteration of software development [12].

Next, this FDA general guideline needs to be rewritten from a software engineering point of view. For this, we need to have a concrete explanation of software usability during software design.

From the software engineering perspective, the usability of software is considered as a non-functional attribute that should be planed for during the development process [15]. Software Usability Engineering defines the usability of software based on seven subjective and objective characteristics: understandability, learnability, memorability, efficiency, low-error rate, compliance to standards and guidelines, and user satisfaction [16, 17]. These software characteristics are evaluated to measure the usability of the final software product.

Most of the existing usability techniques are suitable for the complete software system and do not measure usability in software architecture during development [17]. Based on usability definition, there should be techniques that are capable of assessing the usability of software during the design process. In addition, the usability of software is not limited to user interface design, rather depends on functionality of software such as undo functionality [17].

Moreover, there are sets of design solutions such as usability patterns and usability properties that increase usability of a software application, but these design solutions may cause changes to the software architecture [18]. To consider usability in software architecture design, numbers of architecture sensitive usability patterns are created that can be applied in high-level design such as actions on multiple objects, multiple views, and user profiles [18]. Moreover, there is a software architecture assessment technique called scenario-based assessment technique that provides early assessment of software architecture from usability point of view [18].

4.2 XP and HFE

The FDA requires high quality of usable software to reach HFE in software design. As mentioned before, the FDA suggests following HCI guidelines and usability engineering in software design. The XP is concerned about end users of software product by defining the role interaction designer. Interaction designer as the sub-role of XP programmer is responsible for evaluating usage of the deployed system. This evaluation results in to specify future functionalities of system by defining additional possible user stories. In addition, interaction designer refines the user interface according to usage evaluation which is developed during several iterations to release [7, 19].

During exploration phase, XP does not mention how to deal with usability in the architectural design. To design the software system architecture, XP suggests building system prototype in exploration phase to evaluate possible architectures of software, create the high level design (architecture) of software in exploration and planning phases, and finally the architecture is consolidated in first release [5, 7]. There are two practices in XP that affect the design of architecture: system metaphor and simple design. System metaphors are shared story to describe how the system works and simple design makes easier to understand each design component [20]. But there is nothing in XP about following architectural patterns and assessing usability in architecture.

Constantine and Lockwood [21] believe that XP advocate a user-centered design because of the dependency of XP on customer feedback, setting goals based on what customers want, and getting iterative rapid prototype makes the development team design a system that the customer wishes, but not necessarily what he really needs. One of the drawbacks of this approach is that it is not possible to satisfy all stakeholders when the project has many stakeholders. This is due to the complexity of dealing with multiple stakeholder requests that might be conflicting.

As a result we conclude that XP does not satisfy FDA expectations to provide enough practices to back up usability in software design. Despite the existence of interaction designer, XP practices are not enough to handle usability in software projects that they need considerable amount of design due to scale of project.

4.3 Extending XP to Support HFE

As mentioned earlier, the most important aspect of HFE is usability. Thus, we propose an extension of XP abided by usability inside process. We base our efforts on providing major user stories on exploration phase. Based on the work done by Obendorf et al., they defined sub phases during the XP exploration phase to meet software usability in design [22].

As illustrated in Figure 3, the exploration phase contains following sub-phases before prioritizing stories for each iteration: contextual investigation, requirements scenario, vision/use scenario and stories. Contextual investigation is the method for understanding application domain. This method comes from Contextual

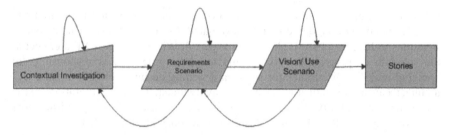

Fig. 3 The extended XP exploration phase

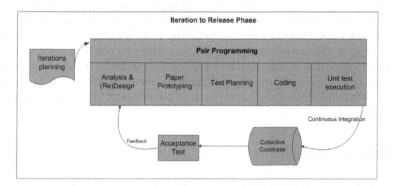

Fig. 4 The Extended XP Iterations to Release Phase

Inquiry with less emphasis on completeness and modeling of gathered information. Contextual investigation uses interview and workplace observation to understand the use context, responsibilities and relationships of end users [23]. Requirement Scenario gets its basics from Problem Scenario in Scenario-Based-Engineering [24]. Requirement scenarios give details about the use of tools and their functionality which are already developed in specified workplace. The Vision or Use Scenario provides a consistent system look with problem statement and its corresponding solution for the specified system [23]. Additional investigation and feedback from users in each increment enhances the requirement scenarios and vision. The first three sub-phases result in to reach XP stories for that increment. In other words, the prioritization of stories for each increment is performed in the exploration phase of the extended XP.

Besides extending the exploration phase in XP to handle usability in design, the extended XP could contain practices such as prototyping and redesigning in the Iterations to Release phase [23].

As mentioned before, in the first iteration during a release, the structural stories are chosen to consolidate architecture. Based on the achieved feedbacks by the end of each iteration, design could be refined or redesigned according to the significance of feedbacks. Since XP design models are informal like drawing on whiteboard, we suggest recording all informal design models to be able to redesign later. In addition, as we mentioned earlier in Section 4.1, using prototyping techniques during design increases the chance of developing usable software. Therefore, by doing paper prototyping after design in XP Iterations to release, the design model could be evaluated for its usability. The extended Iterations to Release phase is showed in Figure 4. We also suggest XP architect as the sub-role of programmer to become the main role who is responsible for applying usability disciplines for story card writing during exploration phase. XP architect can do so with high customer involvement. During exploration phase, this is the responsibility of customer to write use scenario and finally writing story card. In original XP, interaction designer is responsible for evaluating usage of system during iterations to release phase. We suggest interaction designer (XP programmer sub-role) uses paper prototyping as a technique to facilitate usage evaluation of system. As a result of usage evaluation, he is redesigning the already designed features in next iteration according to this evaluation. This redesigning software does not need to change architecture. If XP interaction designer learned that existing architecture does not have the capability for redesign and architecture has to change, this is the responsibility of XP architect to change software architecture to state usability issues. Therefore, we suggest redesigning software features are being done under supervision of XP architect to not violate specified architecture.

5 Related Work

Kent Beck, one of the leading voices of XP, discussed the use of XP for developing secure and safe software in [7]. He points out that features such as safety and security have become the first priority requirements in developing software in areas like avionics and medical systems. He suggests that XP has sufficient capability to support developing such software systems only if additional practices for security or safety are incorporated into XP. He also argues that XP can be adapted to developing software for FDA medical devices by putting the emphasis on the audit process during the XP life cycle. He considers auditing as a continuous practice that starts early in the XP life cycle instead of having it as a separate phase at the end of project. However, it is not explained how XP can be extended to consider the FDA audit process, which is the concern of this paper.

In [25, 26], McCaffery et al. try to address the issue of compliance of medical devices with FDA regulations in process improvement level. For this they suggested the application of a software process improvement process like CMMI can ensure FDA regulatory compliance. Our work differs significantly from their in several respects: our generic framework is not limited to a specific Life-Science regulation. Furthermore, we address the same issue in the level of software development processes by practically extending XP as an important software development methodology.

In [27], Wright explained how he achieved ISO 9001 certification [28] using XP in a software development company. ISO 9001 requires having a quality management framework where business processes of the organization are documented and monitored. Wright proposed a light-weight extension to XP that meets ISO 9001 requirements. He first mapped ISO 9001 process requirements to XP practices. Then, he proposed a way to monitor and measure the process activities. For instance, he created virtual white board to add more features to XP stories and record them. In addition, he related integration, system, and acceptance tests to their corresponding virtual stories. The difference between Wright's work and this paper is that FDA requirements vary significantly from those of ISO 9001. The FDA is concerned with every single activity of a process.

6 Conclusion and Future Work

In this paper, we assessed the ability of XP to meet FDA regulations, which impose stringent requirements on the way software is built. These requirements are in the form of artefacts that a software process must produce for the software system to be FDA-compliance.

Although, we studied the complete set of FDA requirements, we chose in this paper to discuss our mapping process between FDA and XP by focusing on Human Factor Engineering requirements that must be met by any software process that claims to be FDA-compliant. We showed how XP does not support this aspect, and proposed an extension to it.

We intend to work on a larger version of this paper where we discuss every FDA requirement and if and how XP supports it or does not support it. For these activities that are not supported by XP, we intend to propose extension that will consist of adding new roles, practices and artefacts.

References

1. Otto, P.N., Antón, A.I.: Addressing legal requirements in requirements engineering. In: Proc. of the Requirements Engineering Conference, pp. 5–14 (2007)
2. Cugola, G., Ghezzi, C.: Software Processes: a Retrospective and a Path to the Future. In: Proc. of the Software Process Improvement and Practice Conference, pp. 101–123 (1998)
3. Nerur, S., Mahapatra, R., Mangalaraj, G.: Challenges of migrating to agile methodologies. Communications of the ACM 48(5), 72–78 (2005)
4. Qasaimeh, M., Mehrfard, H., Hamou-Lhadj, A.: Comparing Agile Software Processes Based on the Software Development Project Requirements. In: Proc. of the International Conference on Innovation in Software Engineering, pp. 49–54 (2008)
5. Abrahamsson, P., Salo, O., Ronkainen, J., Warsta, J.: Agile Software Development Methods. VTT Publications (2002)
6. Larman, C.: Agile and iterative development: A manager's guide. Addison-Wesley, Boston (2003)
7. Beck, K.: Extreme Programming Explained: Embrace Change, 2nd edn. Addison-Wesley, Reading (2005)
8. Eclipse Process Framework, XP model Library,
   http://www.eclipse.org/epf/downloads/xp/xp_downloads.php

9. FDA Guideline: General Principles of Software Validation; Final Guidance for Industry and FDA Staff (2002),
   http://www.fda.gov/medicaldevices/deviceregulationandguidance/guidancedocuments/ucm085281.htm
10. Glossary of Computer Systems Software Development Terminology,
    http://www.fda.gov/ICECI/Inspections/InspectionGuides/ucm074875.htm
11. FDA Guideline: Design Control Guidance for Medical Device Manufacturers (1997),
    http://www.fda.gov/downloads/MedicalDevices/DeviceRegulationandGuidance/GuidanceDocuments/UCM070642.pdf
12. FDA Guideline: Do It by Design: An Introduction to Human Factors in Medical Devices (1996),
    http://www.fda.gov/downloads/MedicalDevices/DeviceRegulationandGuidance/GuidanceDocuments/UCM095061.pdf
13. FDA Guideline: Medical Device Use-Safety: Incorporating Human Factors Engineering into Risk Management (2000),
    http://www.fda.gov/downloads/MedicalDevices/DeviceRegulationandGuidance/GuidanceDocuments/ucm094461.pdf
14. Boehm, B., Turner, R.: Balancing Agility and Discipline: A Guide for the Perplexed, pp. 26–37. Addison Wesley Professional, Reading (2003)
15. Nielsen, J.: Usability Engineering. Academic Press, Boston (1993)
16. Holzinger, A.: Usability Engineering Methods for Software Developers. Communications of the ACM 48(1), 71–74 (2005)
17. Folmer, E., Bosch, J.: Architecting for Usability; a Survey. The Journal of Systems & Software 70(1), 61–78 (2004)
18. Folmer, E., van Gurp, J., Bosch, J.: Software Architecture Analysis of Usability. LNCS, pp. 38–58. Springer, Heidelberg (2005)
19. Lee, J.C., McCrickard, S.: Towards Extreme(ly) Usable Software: Exploring Tensions between Usability and Agile Software Development. In: Proc. of Agile Conference, pp. 59–71 (2007)
20. Nord, R.L., Tomayko, J.E., Wojcik, R.: Integrating Software-Architecture-Centric Methods into Extreme Programming (XP) Technical Note, CMU/SEI (2004)
21. Constantine, L.L., Lockwood, L.A.D.: Software for Use: A Practical Guide to the Models and Methods of Usage-Centered Design. Addison-Wesley, Reading (1999)
22. Obendorf, H., Schmolitzky, A., Finck, M.: XPnUE – Defining and Teaching a Fusion of eXtreme Programming and Usability Engineering. In: Proc. of HCI Educators Workshop, inventively: Teaching theory, design and innovation in HCI (2006)
23. Obendorf, H., Finck, M.: Scenario-Based Usability Engineering Techniques in Agile Development Processes. In: Proc. of the International Conference on Human Factors in Computing Systems, pp. 2159–2166 (2008)
24. Rosson, M.B., Carroll, J.M.: Usability Engineering: Scenario-based Development of Human-Computer Interaction. Morgan Kaufmann, San Francisco (2001)
25. McCaffery, F., McFall, D., Donnelly, P., Wilkie, F.G., Sterritt, R.: A software process improvement lifecycle framework for the medical device industry. In: Proc. of the 12th IEEE International Conference and Workshops on Engineering of Computer-Based Systems, pp. 273–280 (2005)

26. McCaffery, F., Donnelly, P., McFall, D., Wilkie, F.G.: Software Process Improvement for Medical Industry. Studies in Health Technology and Informatics, pp. 117–124. IOS Press, Amsterdam (2005)
27. Wright, G.: Achieving ISO 9001 Certification for an XP Company. In: Proc. of Extreme Programming and Agile Methods - XP/Agile Universe 2003, pp. 43–50 (2003)
28. The International Standard for Quality Management, ISO,
    http://www.isoqarinc.com/
    ISO-9001-Quality-Management-Standard.aspx

Author Index